2-6-73

Testing of Metals

TESTING OF METALS

Eric N. Simons

South Brunswick and New York
GREAT ALBION BOOKS

© Eric N. Simons 1972

First American edition published 1972 by
Great Albion Books, a division of
Pierce Book Company, Inc.,
Cranbury, New Jersey 08512

Library of Congress Catalogue Card Number: 73-38290

ISBN 0-8453-1138-7

Printed in Great Britain

Contents   1741476

LIST OF ILLUSTRATIONS	page 7
A NOTE ON METRICATION AND TESTING	9
CONVERSION TABLE	10
PREFACE	11
CHAPTER 1: Definitions – The Testpiece – The Tensile Test	13
CHAPTER 2: Tests of Hardness	26
CHAPTER 3: Compression Tests – Tests of Torsion – Impact Tests – Bend Tests – Hard and Brittle Metals Tests – Creep Tests	42
CHAPTER 4: Fatigue Tests – The Dilatometer Test	59
CHAPTER 5: Preparing Testpieces – Recently Developed Tests – Ductility Tests – Formability Tests – Calibrating Test Apparatus	78
CHAPTER 6: Sheet Tests – Testing Bars, Plates and Forms – Testing Non-ferrous Metals – Testing Tubing – The Testing of Wire – Tests on Forged Parts – The Testing of Castings – The Testing of Welds – Testing Miscellaneous Items	98
CHAPTER 7: Radiography – Interpretation – Electron Diffraction – Electron Microscopy – Ultrasonic Testing – Magnetic Crack Detection – Fluorescent Tests – Electrical Testing Methods – Magnetic Analysis – Core Loss Comparison – Current Conduction	112
CHAPTER 8: Corrosion Tests – Fluidity Tests – Tests of Machinability – Tests for Resistance to Oxidation – Tests for Residual Stress – Tests for Season Cracking	

– Stress Corrosion Tests – Tests of Wear – Testing Permanent Magnets – Testing Engine Valves – Lead Corrosion Valve Tests – Gamma Radiography Tests under water – Tests for Molybdenum's Effect on White Iron – Roll Hardness Tests – Die Block Tests – Tests for Steel Damage by Hydrogen – Tests for Stainless Steel Quality – Testing Nodular Iron – Tests of Bolts – Testing Electrodes for Maximum Welding Speed   page 135

CHAPTER 9: Tests of Tool Steels – The Spark Test – Testing Files – Saw Tension Test – Testing Rivets – Testing Sintered Carbides – Testing Composite Metals – Testing Metal Platings – Tests for Solder – Testing Tinplate – Silver Analysis – Electrical Strain Gauges   161

CHAPTER 10: Long Term Creep Rupture Tests – The 'Fastress' Analyser – Gear Testing – Electron Fractography – Closed-Loop Electrohydraulic Tests – Resonant Fatigue Testing – Computerised Testing – Hydraulic Testing Machine – The Modern Metallograph – Hydraulic Fatigue System – The Neutron Radiograph – Portable Hardness Tester on New Principle – New Portable Strain Indicator – Testing New Lorry Cabs – New Penetrants for Magnetic Testing – Portable Ultrasonic Testing Machine – Electron Microscope – In-process Control System – New Type of Extensometer – Tests of Stresses in Aluminium – High Temperature Furnace – The Modern Brinell Testing Machine – Electronic Resistance Weld Testing – Large Scale Notch Toughness Tests – Automatic Non-destructive Testing – Interference Microscopy – A Modern Servo Potentiometer – Xeroradiography – Measuring Non-metallic Inclusions – The Future of Testing   181

ACKNOWLEDGEMENTS   196

BIBLIOGRAPHY   197

INDEX   199

List of Illustrations

PLATES

Tensile Testing Machine 7106 (*W. T. Avery Limited, Birmingham*) page 65
The Olympus PME Metallograph (*Olympus Corporation of America, New York*) 66
Uranium Iron Eutectic 325X under conventional light (*Carl Zeiss Inc, New York*) 83
The same sample of Uranium Iron Eutectic 325X EPI Nomarski Interference contrast (*Carl Zeiss Inc, New York*) 84
Rockwell Hardness Tester, Model 4–JR (*Wilson Instrumental Division, American Chain and Cable Company, New York*) 149
Magnaflux Type KH 12 ac/dc portable magnetic testing unit (*Magnaflux Limited, London*) 150
Marconi Direct Frequency Counter for measurement up to 50MHz, with 0.1μsec timing resolution and autotrigger (*Marconi Instruments Limited, St Albans, Herts*) 167
Creep and Fatigue Testing Machine (*Avery-Denison Limited, Leeds*) 168

IN THE TEXT

1	Typical tensile testpiece before and after the test	17
2	Standard American testpiece for sheet metal	17
3	Typical standard American testpiece for plate metal	17
4	Stress-strain curve	19
5	Load Extension diagram	21
6	Relation of strength of normalised steels to carbon content	24
7	Wrong impression caused by surface flow in Brinell hardness test	31

8	Pyramidal diamond indenter	page 31
9	Vickers test impression	32
10	Hot hardness curve (Vickers diamond test) for a high-speed steel	36
11	Standard Izod impact testpiece, held vertically	48
12	Dilatometer test for a stainless steel (Grenet)	70
13	Explosion bulge test	88
14	Drop weight test	89
15	Main components for an ultrasonic flaw detector	123
16	Direction of crack best detected in modification of eddy current losses	133
17	Sparks of low carbon steel	164
18	Sparks of 18% tungsten high speed steel	164
19	File test curves	170

A Note on Metrication and Testing

At the time of writing, industry and commerce in the United Kingdom are gradually moving towards a standard system of units, but are at present using both metric and SI units, depending on the situation, whereas American workers still use imperial units to a large extent. Clearly, no one system of units can satisfy all readers and it is hoped that the system used in this book—of SI units with imperial units given in brackets—will be considered to be a workable and convenient compromise.

MEASUREMENTS in SI units used in testing are:

Length in metres, symbol m
Mass in kilograms, symbol kg
Time in seconds, symbol s
Thermodynamic temperature in Kelvin, symbol K

SYMBOLS used in the testing of metals include:

b = width
B = plain end diameter
C = plain end minimum length
d = test diameter
E = screwed end size
F = threaded end minimum length
J = energy in Joules (= Newton-metres, Nm)
L_c = minimum parallel length
L_o = gauge length
L_t = approximate total length
r = minimum shoulder radius
S_o = cross-sectional area
t_K = temperature in Kelvin

Conversion Table

	To convert	Multiply by
Area	ft² to m²	$4{\cdot}2903 \times 10^{-2}$
Density	lb/in³ to kg/m³	$2{\cdot}7680 \times 10^{4}$
Energy	ft lb to J	$1{\cdot}3558$
Force	kgf to N	$9{\cdot}8066$
	lbf to N	$4{\cdot}4482$
Length	ft to m	$0{\cdot}3048$
	in to m	$2{\cdot}5400 \times 10^{-2}$
Mass	lb to kg	$0{\cdot}4536$
	tons to kg	$1{\cdot}016 \times 10^{3}$
	tonnes to kg	$1{,}000$
Stress	lb/in² to N/m²	$6{\cdot}8948 \times 10^{3}$
Temperature	°C to t_K	$t_K = T_c + 273{\cdot}15$
	°F to t_K	$t_K = (t_K + 459{\cdot}67)/1{\cdot}8$
Volume	ft³ to m³	$2{\cdot}8317 \times 10^{-2}$
	in³ to m³	$1{\cdot}6387 \times 10^{-5}$

Continental testpieces are similarly standardised, gauge lengths having two standards, 10*d* and 5*d*. To determine the properties of European metals it is advisable to use both lengths at the start of the change to metric units.

Preface

In the testing of metals, times, methods, machines and processes change, and it is some years since the known facts as well as the new have been presented and related to modern problems and conditions. Moreover, I have also endeavoured to make this volume as interesting and comprehensible to the newcomer in this field as to more experienced readers. Consequently, while one cannot avoid the use of technical terms, I have at least tried to make those I have used reasonably clear. It should be noted, however, that when I use the words 'testing speed' in relation to testing, it signifies either the rate at which the jaws or other gripping devices of the testing machine move away from each other, or the rate of load application.

Briefly, the scheme is to introduce the principal forms of testing, dealing first with the mechanical, which are mostly destructive of the testpiece, and secondly, that wide variety of tests known as 'non-destructive'. The following sections cover the testing of various classes of products, individually described, and finally comes an account of a series of innovations, sometimes startling, in modern testing techniques, appliances, machines and researches. These are not grouped in any special order, but the index gives a means of quick reference.

In this last section only have I departed from my rule of adequate explanation of the terms used. Space does not admit of the long statements of theory and practice that would be necessary.

I venture to hope that the book will prove of value and interest not only to the expert, but also to metallurgist, engineer, student, lecturer, teacher, apprentice, works manager, buyer and sales engineer, as well as to ordinary workers in every metallurgical and engineering establishment.

I do not claim direct experience of absolutely all the tests and mechanisms outlined, but my own knowledge, when inadequate, has been supplemented by the help, generously given, of friends and former colleagues, and of large industrial and research associations both at home and abroad. In addition I have drawn upon the compilations of the British Standards Institution, the American Society for Metals, of which I am a Member, the Iron and Steel Institute, etc. I owe special thanks to the magnificent metallurgical section of the Sheffield City Libraries, whose staff have once again given me every possible help. Those organisations that have kindly provided me with photographic prints, drawings and other data concerning their specialised equipment are listed on p 196.

Eric N. Simons
Eastbourne 1971

Chapter 1

The testing of metals can be classified under various headings. For example, it may be regarded as a matter of routine on the one hand and special or experimental research on the other. An alternative subdivision highly popular today is into destructive and non-destructive. Again, it may be grouped into mechanical, chemical, and radiographic tests. Which of these classifications is preferred depends largely on the interests and predilections of the user. This book chooses a different grouping, namely mechanical and metallurgical tests.

Certain considerations must be kept firmly in mind. A test is an experiment carried out on a sample or 'testpiece' of the metal to be studied, and whatever results it gives are peculiar to that testpiece or series of testpieces. Hence, the first requirement in testing is careful choice and treatment of the testpiece to ensure that it corresponds as closely as possible to the metal or metal product whose properties are to be discovered.

To save repeated references to scattered pages, it is as well to begin with a series of definitions of some of the more common aspects of mechanical testing. Such other definitions of technical terms as may be necessary will be given in their proper places.

DEFINITIONS

Compression Test A test to determine the ductility and malleability of metal bars, in which the testpiece, usually 1·5 times the diameter, is compressed to half its length without developing cracks.

Creep Test Tests to determine the ability of a metal to withstand the phenomenon known as 'creep', or continuous deformation under load applied at a specific temperature and at a steady rate, the stress being less than the normal yield strength.

Bend Test A test carried out on a metal sheet or plate to show how far it can be stressed by bending without cracking or fracture. A much rougher test is sometimes carried out on such parts as dredger pins of austenitic manganese steel, which are bent through 180° in the cold state to ensure that their ductility is adequate for their work.

Calibration An experimental determination of absolute values relating to the graduations on an instrument scale.

Fatigue Test A test designed to show the cycles of alternating stress repeatedly applied that can be withstood without failure. It covers stress range, mean stress and cycle number.

Hardness Test One of a wide range of different tests, all to determine the ability of a metal to withstand deformation. The values they give are by no means consistent, but certain relationships exist between them and these will be indicated.

Tensile Test This is the application of an increasing specific load to a testpiece under tension until the metal fractures, to determine the degree of tension it will resist without fracture.

The Object of Testing Usually routine tests are carried out to establish whether a piece or product of metal meets the requirements of a specification. The figures derived from the test not only point to the precise properties to compare with others obtained from a different supplier or with those given by earlier lots from the same supplier, but also help to compare alternative metals. Another object is to provide information in the form of numerical values for the use of designers and engineers, and for works managers to enable them to adapt their processes and techniques to novel products.

THE TESTPIECE

An important point in choosing a testpiece is its location with respect to the main body of metal under test. This is simple when castings have to be tested because the testpiece can be taken from typical castings, or extra pieces of metal can be cast on and cut or machined off, or the entire cast part can be tested, choosing one piece at random from the bulk. The advantages and disadvantages of these three methods will be covered in the specific section devoted to the testing of cast components.

Wrought metal testpieces are not difficult to choose, but there could be variation in the properties they reveal, owing to factors such as the direction of plastic flow in forging and rolling, and the area from which the testpiece is taken, eg the centre, the edge, or from which end, especially in large pieces. The specimens taken from metal forgings suffer from the same advantages and disadvantages as those taken from castings. In general it is good practice to state precisely where the specimen should be taken from and its form and dimensions, if not covered by existing standard specifications (eg BS 18: 1962).

The number of testpieces on which the tests are carried out should be sufficient to ensure that an average can be found for the variations encountered and the values obtained. This is particularly important when the test is to determine the properties of a metal different in composition from any previously used or to determine the effects of new or altered techniques of manufacture. Nowadays testpieces for the determination of mechanical properties are largely standardised, especially the dimensions and form which may greatly modify the data derived from the test unless strictly controlled.

Another cause of variation in test values is a lack of regularity in the preparation of the testpiece, which must be machined and finished in accordance with fixed standards, while the temperature of the testpiece if not controlled could cause distortion or microstructural modifications liable to modify the values given.

Testing Equipment In a later section some types of equipment are described in detail. Here it suffices to say that they usually consist of a machine or mechanism capable of applying a force or load to a testpiece or sample in a particular manner, together with some means of accurately measuring the force in question, and means also of determining exactly the degree to which the testpiece is deformed by the applied load. The apparatus, whatever its type, has to be carefully calibrated from time to time to ensure that wear or other causes have not affected its response.

Preparing the Testpiece The immediate requirement is a testpiece that will give a result as near as possible to the actual performance of the metal under conditions simulated by the test. Hence, it must not incorporate any area known to have been subjected to intense heat, as by a cutting torch, unless a weld is under test. Again, sawing conditions of great friction, of hammering the edges of the testpiece could lead to variations in the properties. Should such areas be present in a testpiece cut from a plate or sheet, the affected metal and also a portion adjacent to it must be cut off, the precise amount removed depending on the size of the piece or part and the chemical composition of the metal. Another influential factor is the type and dimensions of the cutting torch, if this has been used. Specimens cut from sheet and plate by shear blades should usually have about the same amount of metal as the plate is thick removed from the area close to the edge, so that in a plate 6·3500mm thick, there will be a gap of this dimension before the edge of the testpiece is reached.

In testpieces cast on, there is always a danger that harmful stresses may be present, and if this cannot be prevented, the remedy is to give them a stress-relieving treatment according to the type of metal, and by arrangement with the purchaser.

Each type of test demands its own form of testpiece. The commonest and most easily understood mechanical test is the tensile test, in which a testpiece milled to form is pulled in opposite directions, the load gradually increasing until the testpiece breaks. The form of a typical testpiece is shown in Fig 1 from which it is seen that both ends have a shoulder, A, fixed in the machine. The real testpiece is the portion marked C, ie the parallel middle portion between the two punch marks indicating the gauge length. The connecting pieces, B, between middle and shoulders have an intermediate diameter, and there must be no abrupt changes of cross section between B and C. This is achieved by means of a sufficiently large radius between the connecting parts. Any sharp change of cross section causes local concentrations of stress and possibly the fracture of a testpiece outside its gauge length. British Standard Specifications are insistent that if a tensile testpiece fractures outside the middle of half its gauge length, the test may be scrapped and another made. The ends of the testpiece are formed according to the gripping facilities available on the machine.

Some testpieces are screw-threaded at the end for better gripping during the test, and here too the portions between ends and parallel section have to be given a good radius. In testing thin

THE TESTPIECE

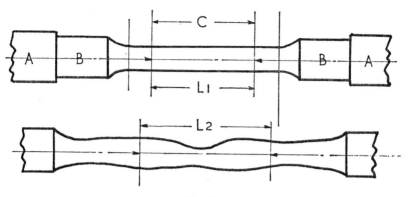

FIG 1

sheet and strip, radius is of major importance. The testpiece must be well finished, the surface within the gauge length being smooth, this finish being obtained by using a grade 00 emery paper or a smooth abrasive wheel.

The form of testpieces for sheet, plate, strip and tubing with thin walls is different from that used for cast and forged parts, and Fig 2 shows the sheet metal testpiece specified in the United States, while Fig 3 shows the testpiece used for plate in the same

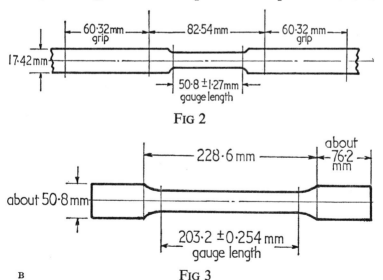

country. The round testpieces need less preparation, a smaller amount of metal, and usually give more consistent results. In addition to the plain and screwed ends already mentioned, some have shouldered round ends to be gripped in vee-type wedge grips for use in special fixtures. The most satisfactory results with each type are obtained when the seats of the tension members are spherical. Round testpieces are usually machined by lathe, either individually or in quantity, with a suitable automatic machine tool such as a turret lathe, or alternatively a screw-cutting lathe operated automatically if the quantity required warrants it.

Where machine tools using tungsten carbide cutting-tools or those of super-high-speed steel are not available, the required finish can be given by grinding, but in this case the wheel must be chosen with the greatest possible care to prevent grinding cracks on the surface caused by heat generation, while the proper speeds, feeds and cooling medium must also be carefully chosen. Some slender testpieces of particular materials are liable to warp or distort slightly when hardened, so that if the data are required on a hardened sample, it may be advisable to leave a little additional metal on, which can be machined away to remedy this distortion without affecting the finished dimensions of the testpiece.

The screw-threaded ends of testpieces have to be concentric with the testpiece axes. Testpieces with their ends shouldered have to have plane inner or loading surfaces normal to the testpiece axis; the ends of plain round testpieces must be symmetrical to and concentric with the testpiece axis. The holding devices for the ends should provide uniform and axial stress in all three types.

We shall deal with the preparation of other testpiece forms in later sections.

THE TENSILE TEST

In this test a specially prepared testpiece is increasingly loaded by being pulled at each end in opposite directions until it breaks. If during the test the precise increase in length or extension relating to each increase of load is carefully measured and the

THE TENSILE TEST

results plotted, a curve of the type shown in Fig 4 is obtained (often wrongly termed a stress-strain diagram), and gives a great deal of useful practical information. The straight line OA, termed the line of proportionality, is indicative of the phenomenon defined by Hooke's Law, namely that for perfectly elastic bodies stress is proportional to strain. The point A marking the end of the proportionality range is termed the 'elastic limit', because under any lower starting load the metal behaves as if it were truly elastic, recovering its original dimensions as soon as the load is taken off. From this portion of the curve Young's modulus of elasticity, Y, can be ascertained, and is defined as the ratio of longitudinal stress to longitudinal strain. For example, if a testpiece originally of length L and cross-section S_o is elastically stretched an amount e by a load P acting along its axis, an equation is obtained:

$$Y = \frac{\text{longitudinal stress}}{\text{longitudinal strain}} = \frac{P/S_o}{e/L} = \frac{PL}{S_o e}$$

In Fig 4 it is plain that if the load is increased beyond A, proportionality stops and the curve descends. Removal of the load reveals that the testpiece does not fully regain its original dimensions, and assumes a specific, but small, degree of plastic or

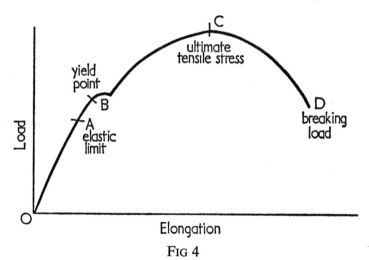

FIG 4

permanent deformation. In the end, point B is reached, and here quite a small increase in load causes a comparatively large extension, meaning that the metal has suddenly given way under the load. This point shows the yield point of the material, and from this a most useful indication of its toughness or strength can be obtained.

In machines that test by lever, the yield point is frequently revealed by a sharp falling of the beam, or in hydraulic machines by a stationary position, or even a backward movement, of the needle on the load indicator. The yield point is therefore the minimum load causing the testpiece to extend without a further increase of load. On occasion the load-extension curve indicates a fall in the stress when the yield point, is attained.

With increase of load beyond the yield point, the testpiece slowly stretches or creeps until a point is reached at which the cross-section will support no additional load. This is termed the ultimate tensile or maximum stress, and is indicated by C in Fig 4. Beyond this point the testpiece, which up to now has remained largely parallel, betrays a narrowing of the neck, that is, in some places the cross-section quickly decreases, so that only a minute addition to the load is sufficient to produce further extension of a highly plastic type, and when this has reached its limit, the metal fractures. The load, point D in Fig 4, is that at which fracture occurs, and is termed the breaking point, but as it has no genuine significance, it is usually omitted from the test report.

The properties of elastic limit, yield point and ultimate tensile stress are a guide to the toughness or strength of the metal tested, and its ability to withstand an applied stress. For a particular metal the elastic limit declines with the sensitivity of the extension measuring device, and it is rare for this value to be embodied in modern specifications.

Proof Stress Many authorities hold that yield point is a more important indication of the toughness of a metal than the ultimate tensile strength. It has to be noted, however, that the majority of the non-ferrous alloys and a considerable number of high tensile steels do not show a definitive yield point, their load extension curves being of the type shown in Fig 5. To give the practical engineer a proper basis for his work, the proof stress has been developed. This is the load or stress that, applied for a minimum period of 15min, gives a plastic extension or permanent set of 0·1% (or 0·5% with certain alloys) of the original gauge length as measured by an extensometer.

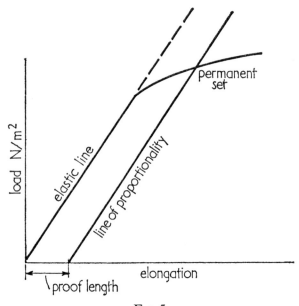

FIG 5

Various methods are adopted to ascertain the proof stress, the obvious one being to load and unload the testpiece at successive increases of load until eventually the predetermined permanent set is obtained. This, however, is a tedious method, the required value being more conveniently obtained from an accurately-drawn load-extension diagram, as in Fig 5. The given proof length (0·1 or 0·5% of the gauge length) at which this line cuts the load extension curve indicates the proof stress value, since the difference between the total extension shown by this point and the corresponding point on the theoretical elastic line obviously corresponds to the previously established permanent set. Proof stress elongation varies for different metals.

Elongation If the two fractured ends of the testpiece are carefully fitted together, the final extension over a specific gauge length is determined. If L_1 and L_2 in Fig 1 are the original and final distances between two punch marks on the testpiece, the percentage of elongation is evaluated as follows:

$$\text{Elongation (\%)} = \left\{\frac{L_2 - L_1}{L_1}\right\} \times 100.$$

The value so arrived at measures the metal's ductility, but it is essential to specify the gauge length in each instance, because the total extension is the sum of two factors, namely, the general extension over the entire length of the testpiece, and the local extension confined to the area close to the break. It is demonstrable that the local extension is a constant for metal of a specific diameter, as is the general extension for each unit of gauge length. Since this latter is usually much less than the local extension, it is plain that the increase of gauge length decreases the calculated value for the elongation percentage. In consequence, testpieces have to conform to standard dimensions.

British Standard Testpieces In Britain the British Standards Institution has standardised testpieces (BS 18: 1962) by the expression $L_o = 4\sqrt{S_o}$ where L_o is the gauge length and S_o the cross-sectional area. The commonest of their specifications for both ferrous and non-ferrous alloys specifies a gauge length of 50·8mm (2in) and a diameter of 19·95mm (0·7854in), giving a cross-sectional area of $5d$, which dimensions obviously meet the above relation.

Reduction of Area The elongation of the testpiece is necessarily accompanied by a reduction in cross-sectional area. By measuring the smallest diameter after fracture, the remaining item obtainable from the tensile test (the reduction of area %) may be calculated by the equation: Reduction of Area %

$$= \frac{(\text{Initial Area} - \text{Area after Fracture})}{\text{Initial Area}} \times 100.$$

This reduction figure is valuable because it indicates the ductility and toughness of the metal under test. Different metals, or even the same one under different conditions, may show almost

THE TENSILE TEST

identical values for the elongation, but the reduction of area values may differ considerably. It is significant that in general a good reduction of area is associated with a comparatively high shock resistance as measured by the impact test (see page 47). In considering tensile test results, the important factor is the combination of the different items, and for most engineering purposes an extremely tough but adequately ductile metal is the objective. This is often achieved by mechanical or heat treatment only, since cast metals usually give low elongation and reduction of area values.

It is even possible to use the tensile curve as a means of calculating the stress-strain curve in torsion, if the fully corrected curve is used. In every range of strain common experience reveals that the torsion points decline by the application of the shear-strain-energy hypothesis and that of maximum shear stress. The difference between the two extremes is only 15%, according to Gensamer.

Carbon Content Tensile properties are influenced by modification of the carbon content in steel etc. As soon as carbon enters into iron, the tensile strength in the 'normalised' condition rises from about $308 \cdot 7 MN/m^2$ (20 tons/in²) up to about $926 \cdot 1 MN/m^2$ (60 tons/in²), the higher figure corresponding to a carbon content of about $1 \cdot 0 \%$. Addition of carbon if continued does not increase the tensile strength, but causes it to fall. It is known that $0 \cdot 82\%$ carbon is the eutectoid or pearlitic microstructure, and this figure is close enough to $1 \cdot 0\%$ to show that a potential relation exists between the eutectoid point and the tensile strength of the steel.

Any increase in the tensile strength caused by adding carbon is accompanied by a decline in ductility, since the two properties cannot coexist, so that as steel is higher in tensile strength, its hardness increases. Fig 6 shows the relation between the tensile properties of 'normalised' steels and their carbon content.

Crow's 4-point Method British Standard testpieces (BS 18: 1962; BS 1452: 1961; BS 1367: 1947) have the central portion of the cast testpiece machined to standard dimensions of usually $0 \cdot 7854 d^2$ cross section by $5 \cdot 65 \sqrt{S_o} = 5d$ gauge length. The elastic extensions of the testpieces are measured by highly sensi-

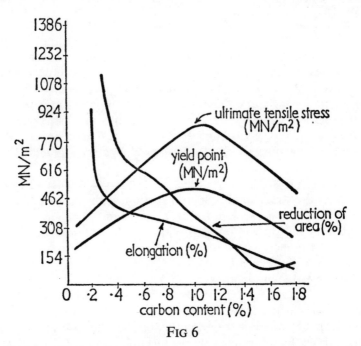

Fig 6

tive instruments termed 'extensometers'. Proof stress is now determined by means of a four-point method evolved by C. A. Crow, which does away with the tedious plotting of curves for each specimen tested, but calls for skill and experience in its use. A proof stress indicator is now obtainable in which an instrument with a cursor and a transparent arm, combined with a range of charts for the various metals, directly indicates the precise proof stress without any need for calculation.

Continental testpieces are also standardised, and gauge lengths have two standards, $10d$ and $5d$ respectively. It is usually advised that in testing to determine the properties of metals obtained from Europe, both lengths should be used, at all events at the outset when changing over. Conversion data are obtainable from BS Handbook No 13, 1951, p 30.

The way in which a tensile test is carried out has been shown to influence the values given, especially where yield values are concerned. These may be up to 10% higher, with a corresponding decrease in elongation if either of the above factors is increased. The ultimate strength is not greatly affected, but compression strength figures may be considerably modified. Hence, standard methods of tests are always advisable, and careful record should be made and kept of the speed employed.

Chapter 2

TESTS OF HARDNESS
Hardness is a loose term, for its meaning depends much on the person using it. Broadly, it is the ability of a metal to resist plastic deformation, normally achieved by indentation with a substance more resistant, that is, harder, than itself. The term is, however, so wide that it includes rigidity, and resistance to scratching, abrasion and cutting action.

Inevitably, therefore, a variety of testing methods exists. In Britain great use is made of the *Brinell hardness test*, in which a small hardened and tempered chromium steel or tungsten carbide ball of 10mm diameter is forced by pressure into the metal to be tested. The testpiece is placed on the platform of the testing machine either universal or specially designed, the platform being elevated by screw to make contact with the ball. This achieved, an oil pump applies gradually increasing pressure, which is read from a manometer or pressure gauge fixed to the machine. A regulating balance is also fitted, the centre pin of which rests on the balls in the cylinders. These act as a piston and carry weights, rendering the pressure variable with hardness. The load is known in advance, and is usually 3,000 kgf, under which load the weights float.

The pressure is applied for 10s and then released, the machined and fine emery polished testpiece being removed and taken to a microscope whose inner lens has a millimetre scale divided into tenths, so that the diameter of the indentation made by the ball can be measured with accuracy. Non-ferrous metals are mostly subjected to a load of 500 kgf for 30s. The Brinell number, corresponding to the hardness of the metal is obtained from the equation: $BN = \dfrac{P}{\text{Spherical Area of Impression (mm)}}$,

BN being the number, P the load (kgf). An alternative and more precise equation is $\dfrac{P}{\dfrac{\pi D}{2}(D-\sqrt{D^2-d^2})}$, d being the diameter of the ball impression, and D the diameter of the ball. A third and simpler equation is $BN = \pi D^P \delta$, δ being the depth of the ball impression.

It will be apparent that the harder the metal, the smaller the indentation diameter, and consequently the higher the Brinell number. For thin metals, however, the load of 300 kgf is too heavy, assuming the relation of indentation depth is D^2 is constant. The ball and load are, therefore, variable and the load is applied slowly for 30s.

It is easier to measure the indentation diameter than the depth, and from the properties of the circle we arrive at the equation $\left(\dfrac{d}{2}\right)^2 = (D-h)h$, or $(D-\delta)\delta$ so that $h^2 - Dh + d^2 = 0$.

This being a quadratic equation in h, its roots 4 are $h = \dfrac{D + \sqrt{D^2 - d^2}}{2}$, h being the impression depth. Rejecting the value derived by taking the $+$ sign, we obtain the value of the ball outside the indentation, and achieve the previously mentioned equation $BN = \dfrac{\text{Load }(P)\text{ in kgf}}{\dfrac{\pi D}{2}(D - \sqrt{D^2 - d^2})}$

When the load is varied in standard machines, the variation is by increases of 500kgf each. After application of the load to the softer metals for 30s minimum, the pressure is released and the testpiece removed by lowering the table or platform. The impression diameter is measured by a low power microscope with a graduated scale in the eyepiece. The middle of the indentation should be more than 2·5 × diameter from the testpiece edge, and the thickness more than 7 × indentation depth. The indentation must not be visible on the other side of the thin testpiece.

The Brinell number is largely indicative of the metal's tensile strength, which can be calculated by multiplying the number by a constant according to the properties of the metal. (See Table 1 below from BS 240, Part 1, 1962.)

In the Brinell machine, the ball diameter is fixed, but there are other machines based on the same principle in which the ball has a different diameter. For example, small hardness testing machines on the Brinell principle may be of 1, 2 or 5mm diameter with loads ranging from 10 to 50 kgf. Random modification of either ball diameter or loads may considerably affect the Brinell number for a specific metal, since there is no geometrical similarity between the indentations and the load/unit of volume displaced is not constant.

It would, therefore, have been better if the hardness scale in this test had been obtained by dividing the load by the spherical surface instead of by the impression volume. The hardness values thereby obtained would then have represented specific work of indentation and a great deal of argument and misunderstanding would have been avoided. When the ratio of load to the ball diameter squared has a constant value, geometrically similar indentations are obtained, so that the hardness numbers for the same material are identical. This presupposes a normal distribution of constituents in the microstructure. To achieve this identity of hardness numbers, a load of 30kgf and a 1mm ball, or 750kgf and a 5mm ball, or 3,000kgf and a 10mm ball, are used.

Ratios now specified by the British Standards Institution are:

L/D^2	Material
30	Ferrous metals of similar hardness
10	Copper alloys
5	Commercially pure copper
1	Lead and tin alloys

Ball hardnesses are also specified in BS 240, Part 2, 1964.

Safeguards are essential in hardness testing to ensure that results are accurate. In the first place the testpiece thickness should, as earlier stated, exceed $7 \times D$ for hard metals and $15 \times D$ for soft metals. On occasion the indentation is not fully spherical, and then the maximum and minimum diameters are

TESTS OF HARDNESS

TABLE 1
BRINELL'S HARDNESS NUMBERS

Diameter of Steel Ball   10mm
Pressure   3,000kg

Diameter of Ball Impression	Hardness Number	Calculated Tonnage	Diameter of Ball Impression	Hardness Number	Calculated Tonnage	Diameter of Ball Impression	Hardness Number	Calculated Tonnage
mm			mm			mm		
2	946	206	3·70	269	59	5·35	124	28·5
2·05	898	196	3·75	262	57	5·40	121	28
2·10	857	187	3·80	255	55	5·45	118	27
2·15	817	178	3·85	248	54	5·50	116	26·5
2·20	782	171	3·90	241	52	5·55	114	26
2·25	744	162	3·95	235	51	5·60	112	25·5
2·30	713	155				5·65	109	25
2·35	683	149	4	228	50	5·70	107	24·5
2·40	652	142	4·05	223	49	5·75	105	24
2·45	627	136	4·10	217	47	5·80	103	23·5
2·50	600	131	4·15	212	46	5·85	101	23
2·55	578	126	4·20	207	45	5·90	99	22·75
2·60	555	121	4·25	202	44	5·95	97	22·5
2·65	532	116	4·30	196	43			
2·70	512	112	4·35	192	42	6	95	22
2·75	495	108	4·40	187	41	6·05	94	21·5
2·80	477	104	4·45	183	40	6·10	92	21
2·85	460	100	4·50	179	39·5	6·15	90	20·75
2·90	444	97	4·55	174	39	6·20	89	20·5
2·95	430	94	4·60	170	38·5	6·25	87	20
			4·65	166	38	6·30	86	19·75
3	418	91	4·70	163	37·5	6·35	84	19·25
3·05	402	88	4·75	159	36·5	6·40	82	19
3·10	387	84	4·80	156	36	6·45	81	18·75
3·15	375	82	4·85	153	35	6·50	80	18·5
3·20	364	79	4·90	149	34	6·55	79	18·25
3·25	351	76	4·95	146	33·5	6·60	77	17·75
3·30	340	74				6·65	76	17·5
3·35	332	72	5	143	33	6·70	74	17
3·40	321	70	5·05	140	32	6·75	73	16·75
3·45	311	68	5·10	137	31·5	6·80	71·5	16·5
3·50	302	66	5·15	134	31	6·85	70	16·25
3·55	293	64	5·20	131	30	6·90	69	16
3·60	286	62	5·25	128	29·5	6·95	68	15·75
3·65	277	60	5·30	126	29			

measured, the mean value being taken to establish the hardness. This will not provide a fully accurate figure, but one close enough for normal requirements in practice. The testpiece or area should be a minimum of $2\frac{1}{2}D$ wide, with a flat and smooth surface. For the smaller balls of diameter below 5mm the testpieces should be made smooth by rubbing with grade 00 emery paper, but not buffed. The absence of such precautions may result in undesirable fluctuation in the values obtained. The microstructure of the metal must be uniform. Minute amounts of retained austenite (a microstructural constituent), soft skin (decarburisation) and similar features which may not always be revealed by the standard Brinell test can notably affect wear resistance.

In any event the test is not wholly trustworthy, and the numeral shown may be and on many occasions is lower than the true hardness of the metal being tested when that metal is hard. Extremely hard metals, in fact, sometimes deform the ball, so that indications of more than 550 Brinell are never trustworthy. The use of a tungsten carbide ball extends the range considerably, however. The test is not affected by vertical motion, but is unsuitable for metals at high temperature. Another drawback is that metals of more than usual hardness may split under the load. There are also possibilities of error arising from faulty readings of hand and edge operation.

The Brinell test is, therefore, mostly limited to large testpieces or parts. It can be used on finished parts without damage, except where the indentation mark is undesirable. Fractures can be examined for hardness in those instances where the tensile test would be impossible. Additional uses are the influence of cold working, and a rough evaluation of ultimate tensile strength by means of a table such as Table 1.

The Vickers Diamond Test Because of the limitations of the Brinell test, it was found desirable to replace the ball by a pyramidal form of diamond. When the Brinell machine is put to work on soft metal, the surface of the metal is elevated by plastic flow about its initial height, as shown in Fig 7. This means that the indentation diameter d' is greater than it should be, so that the hardness numeral is lower than the true value. Deformation

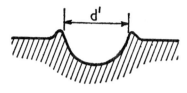

FIG 7

of the ball by extremely hard metals is another source of error. The square pyramidal diamond (Fig 8) indentation is studied under a microscope and regulated to ensure that one corner just makes contact with a fixed knife-edge that can be seen through the microscope eyepiece. A movable knife-edge is then adjusted to make contact, equally slight, with the corner diagonally opposite. By turning a knob a counter is set in motion and the figures it indicates on the arc are transformed into hardness values by means of tables.

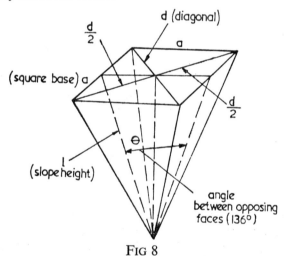

FIG 8

The surface area of the impression is $2al$, a being the side of the square base and l the slant or slope height, but in use the diagonal d of the base is more easily and precisely read than the side. In consequence we get $d^2 = a^2 + a^2$, so that $a = \dfrac{d}{\sqrt{2}}$ and

$\frac{a}{2} = l \sin \theta$, ie $l = \frac{a}{2 \sin \theta}$. In this equation θ is the semi-vertical angle between opposing pyramidal faces. By substitution of these values we have surface area $= \frac{d^2}{2 \sin \theta}$, the hardness numeral being thus provided by the expression DPN (Diamond Pyramid Number) $= \frac{2l \sin \theta}{d^2}$.

The diamond used is 136° between opposite faces, and all impressions are geometrically similar, so that different angles of penetration do not cause variations. The load can be 1–5kgf in stages of 1kgf, or 5–10kgf in stages of 5kgf, depending on the hardness and thickness of the metal.

In looking through the eyepiece of the self-centring, angular microscope, the metallurgist makes the left-hand end of the diagonal coincide with the fixed knife edge, using a pair of focussing screws. The counter is rotated by means of a knurled head, being located on the side of the ocular, the movable eyepiece being inside the instrument. Correct focussing and adjustment result in a view of the indentation as shown in Fig 9.

FIG 9

If the ocular indication is multiplied by 0·0508 it provides the length of the diagonal in mm, so that from this the surface area of the pyramid may be calculated and the conversion tables used (see Table 2) and BS 427, Part 1, 1961. The load is applied by levers and ranges from 1 to 100kgf, but the hardness numbers do not depend on the load. The period of impression is automatically regulated by a signal device in co-operation with a pointer on the machine side. The load itself is governed by the degree of hardness and the microstructure of the testpiece.

Where the metal is of coarse grain and heterogeneous, it is best to apply a large load, but small loads are better for soft metals whose microstructures are of small grain size and uniform. Polishing of the surface with 00 grade emery paper is also necessary but buffing is undesirable.

The load on the plunger on which the diamond is mounted is applied by cam, revolving on a drum round which is coiled a wire carrying a weight. The hardness numerals are much the same as the Brinell up to about 300, but are more accurately representative of the true hardness above this figure, and in Britain are expressed as VPN (Vickers Pyramid Number). An alternative equation for these is $\left\{0.4636 \times \dfrac{P}{R^2}\right\} \times 10^6$ for a 16·93mm (⅔in) objective, where P is the load in kgf and R the ocular reading in mm. Thus, VPN = $\left\{0.07418 \times \dfrac{P}{R^2}\right\} \times 10^6$ for a 38·1mm (1½in) objective.

The test is unquestionably a noteworthy advance in testing metal hardness by indentation. It is equally precise for materials either thin or thick, hard or soft. Any degree of hardness can be determined, and the vertical motion does not affect the indications. That loading is automatically applied means that there are no errors in the loading rate, time and impact. The indentation, being square, is read with less difficulty, but small test-pieces have to have a higher polish than is necessary for those in the Brinell test. The time required for ocular regulation is less, and mounted specimens can be used. Large specimens can be tested if an extension is employed, and it is also possible to determine the hardness of metals at high temperatures up to 900° C (1,650° F).

Hot Hardness Tests These tests of metal hardness at high temperatures are known as hot hardness tests, and are useful in providing some idea of how a tool or die will behave when its use results in the attainment of these temperatures. Fig 10 shows the hot hardness curve given by a Vickers diamond test on a high speed steel tool, hardened at 1,320° C (2,410° F), and tempered at 570° C (1,060° F). The steel contained 22% tungsten

TABLE 2
CONVERSION TABLE OF VARIOUS HARDNESS NUMBERS

This table applies only to chemically and mechanically uniform steels containing carbon, chromium, nickel, vanadium, molybdenum, silicon and manganese, none of the alloys exceeding 4 per cent. No cold-worked steels are included. All testing methods used were those recommended by the manufacturers of the respective instruments.

Vickers or Firth	Brinell			Rockwell								Monotron		Drop Hardness mm/kg	Herbert Pendulum		Scleroscope	Duroskope	Moh
	Diameter mm	Standard Ball	Tungsten Carbide Ball	C 150kg	D 100kg	A 60kg	B 1/16in Ball	E 1/8in Ball	30-N	30-T	Durometer	Cons Dia	Diam Brin		Time S	Time D			
1,224	2·20	780	872	72	78	84					87	28	130	1,200			90	64	106	8·5
1,116	2·25	745	840	70	77	83					86	31	122	1,130			85	59	102	8·5
1,022	2·30	712	812	68	75	82					84	32	111	1,030			80	56	98	
941	2·35	682	794	66	74	81					83	34	103	956			76	52	94	
868	2·40	653	760	64	72	80					81	37	96	894			72	49	91	54	
804	2·45	627	724	62	71	79					79	40	91	850			67	47	87	53	8·0
746	2·50	601	682	60	70	78					78	43	86	804			63	45	84	52	
694	2·55	578	646	58	68	78					76	45	82	767	1,400	60	42	81	51	
650	2·60	555	614	56	67	77					74	46	78	727	1,300	56	40	78	50	7·5
606	2·65	534	578	54	65	76					72	48	74	690	1,225	53	38	76	49	
587	2·70	154	555	52	64	75					70	49	71	660	1,160	51	37	73	49	
551	2·75	495	525	50	63	74					69	50	68	630	1,095	48	36	71	48	
534	2·80	477	514	49	62	74					68	51	66	610	1,050	47	35	68	48	
502	2·85	461	477	48	61	73					67	52	63	586	1,005	44	34	66	47	
474	2·90	444	460	46	60	73					66	53	61	566	950	41	33	64	46	7·0
460	2·95	429	432	45	59	72					65	55	59	548	910	30	32	62	46	
435	3·00	415	418	43	58	72					64	56	57	530	880	37	30	61	45	
423	3·05	401	401	42	57	71					63	58	55	510	840	35	29	59	45	
401	3·10	388	388	41	56	71					62	59	53	490	810	34	28	57	44	
390	3·15	375	375	40	56	70					61	60	51	471	780	33	27	56	44	6·5
380	3·20	363	364	39	55	70					60	62	50	462	750	32	26	54	43	
361	3·25	352	352	38	54	69					59	63	48	453	725	30	26	53	43	
344	3·30	341	341	36	53	68					58	64	47	433	700	29	25	51	42	
334	3·35	331	330	35	52	67					57	65	45	414	675	28	24	50	42	
320	3·40	321	321	33	50	67					56	66	44	408	650	27	24	49	41	
311	3·45	311	311	32	50	66					55	68	42	390	630	26	23	47	41	
303	3·50	302	302	31	49	66					54	69	41	380	610	24	23	46	40	6·0
292	3·55	293			30	48	65				53	70	40	370	590	24	22	45	40	
285	3·60	285			29	47	65				52	71	39	360	570	23	22	44	40	
278	3·65	277			28	47	64				51	72	38	350	550	23	21	43	39	
270	3·70	269			27	46	64				50	73	37	340	535	22	21	42	39	
261	3·75	262			26	45	63				49	75	36	331	520	22	20	41	38	
255	3·80	255			25	45	63				48	76	35	322	505	21	20	40	38	
249	3·85	248			24	44	62				47	77	34	313	490	21	19	39	37	5·5
240	3·90	241			23	43	62	102		46	85	78	33	304	475	20	19	38	37	
235	3·95	235			21	42	61	101		45	84	79	32	295	460	20	19	37	37	

Table 2—continued

Vickers or Firth	Brinell Diameter mm	Brinell Standard Ball	Brinell Tungsten Carbide Ball	Rockwell C 150kg	Rockwell D 100kg	Rockwell A 60kg	Rockwell B 1/16in Ball	Rockwell E 1/8in Ball	Rockwell 30-N	Rockwell 30-T	Durometer	Monotron Cons Dia	Monotron Diam Brin	Drop Hardness mm/kg	Herbert Pendulum Time S	Herbert Pendulum Time D	Scleroscope	Duroskope	Moh
228	4·00	229		20	41	61	100		44	83	81	31	286		450	19	18	36	36	
222	4·05	223		19	40	60	99		43	82	82	30	277		440	19	18	35	36	
217	4·10	217		17	39	60	98	110	42	82	83	29	268		430	19	18	34	36	
213	4·15	212		15	38	59	97	110	40	81	84	29	268		415	18	17	34	35	
208	4·20	207		14	37	59	95	110	39	81	86	28	259		405	18	17	33	35	
201	4·25	201		13	37	58	94	109	38	80	87	28	259		395	18	17	32	35	
197	4·30	197		12	36	58	93	109	37	79	88	27	250		385	17	16	31	34	
192	4·35	192		11	35	57	92	108	36	78	89	26			375	17	16	30	34	5·0
186	4·40	187		9	34	57	91	108	35	78	90	26			365	17	16	30	34	
183	4·45	183		8	34	56	90	108	34	77	92	25			360	17	16	29	33	
178	4·50	179		7	33	56	90	107	33	77	93	25			350	17	16	29	33	
174	4·55	174		6	33	55	89	107	32	76	94	24			340	16	15	28	33	
171	4·60	170		4	32	55	88	106	31	76	95	23			335	16	15	28	32	
166	4·65	167		3	32	54	87	106	30	75	96	23			330	16	15	27	32	
162	4·70	163		2	31	53	86	105	29	74	97	22			320	16	15	27	32	
159	4·75	159		1	31	53	85	105	28	73	99	22			315	16	15	26	31	
155	4·80	156		0	30	52	84	104	27	73	100	21			310	16	15	26	31	
152	4·85	152					83	104		72	102	21			300	16	15	25	31	
149	4·90	149					82	103		71	104	20			295	15	14	24	30	4·5
146	4·95	146					81	103		71	105	20			290	15	14	24	30	
143	5·00	143					80	102		70	106	19			285	15	14	24	30	
140	5·05	140					79	102		69	108	19			280	15	14	23	29	
138	5·10	137					78	101		68	109	19			275	15	14	23		
134	5·15	134					77	101		68	111	18			270	15	14	23		
131	5·20	131					76	100		67	113	18			265	14	13	22		
129	5·25	128					75	100		67	115	18			260	14	13	22		
127	5·30	126					74	99		66	116	17			255	14	13	22		
123	5·35	123					73	99		66	117	17			250	14	13	21		
121	5·40	121					72	98		65	119	17			245	14	13	21		
118	5·45	118					71	98		64	121	16			240	14	13	21		
116	5·50	116					70	97		63	122	16			235			20		
115	5·55	114					68	97		63	123	16			233			20		
113	5·60	111					67	96		62	124	15			230			20		
110	5·65	110					66	95		61	126	15			228			20		
109	5·70	109					65	95		60	127	15			225			19		
108	5·75	107					64	94		59	128	14			223			19		

and 12 per cent cobalt. The curve indicates that the steel keeps its hardness up to about 550° C (1,020° F).

The method of testing is to use an electric furnace for heating up the steel and holding the temperature steady. Nitrogen gas is forced through the furnace, after which the testpiece is introduced, and as soon as it has attained the furnace temperature, the hardness test is carried out on the spot, even though the indentation will be measured at room temperature, this causing no appreciable error.

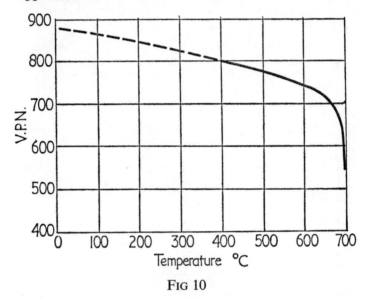

FIG 10

The Firth Hardometer Test This is broadly similar to the Vickers diamond hardness test, but the machine is less complicated. The load is either 10, 30 or 120kgf and is applied by the compression of calibrated helical springs, so that inertial influences are eliminated.

The Rockwell Hardness Test Like the Vickers machine, this employs a diamond, but one conical in form, though in both Vickers and Rockwell machines balls can be used if desired. The ball in these instances is 1·588mm ($\frac{1}{16}$in) diameter. However, the modern tendency is to use the diamond cone (angle 120°), and though machines are graduated in two scales, C and B, most values are given on the C scale for the harder metals. There are also more scales, namely A, D, E and F, each representing a different combination of indenter and load. A recent machine of British manufacture based on the same principles is the Avery hardness tester, which is an improvement on the Rockwell as regards the means of measurement and in certain other respects.

The hardness numeral is obtained by measuring the *depth* of the impressions superimposed on one another. The first impression made is produced by a load of 10kgf, termed the preliminary, which penetrates the superficial layers of the steel or other metal. This minor load is applied and is indicated on a small subsidiary dial at a point marked 'set'. The primary pointer is then set at zero by revolving the inclined edge of the dial until it shows zero. The load is now raised to 150kgf (140kgf is added to the 10kgf of the initial load) by button pressure. The 140 kgf load is then removed by a side lever and the hardness number is shown on the instrument dial, which is marked off in graduations from 0 to 100, representing a single rotation of the pointer and a penetrator movement limited to 0·02mm. In the Rockwell testing machine the diamond scale load limit is 10 to 140kgf.

The test takes only about 10s, which means that it is ideal for routine work, and as the indentation is almost invisible to the naked eye, no particular harm is done to the finish of the work. The surface concerned needs no other preparation than the elimination of oxide. On the other hand, the machine's scale is not nearly so open, so that errors may occur in some instances from misreading of the scale rebound, while should there be any modification in the form of the penetrator face, this may also impair the accuracy of the readings.

To ensure that the maximum accuracy is obtained, therefore, certain precautions are essential. The machine should be so positioned that vibration has no serious disturbing effect on the dial gauge readings. When the minor load is being used, the large-scale needles on the dial gauge must be 'set' within + or − 5 scale divisions. It may happen that as the platform of the machine is raised, the needle stops, though it does not do so on 'set'. It is not then necessary to adjust the lifting screw again, the dial being rotated so that 'set' comes under the needle.

Assuming the machine is of orthodox pattern, the dashpot speed in giving it the major load should allow the crank to finish its traverse in 5s when no testpiece is in the machine, the machine being set up to give 100kgf load. Similarly the surface testing machine is set for 7s to give a major load of 30kgf.

In applying the major load there should be no hand pressure on the crank after release of the lever system. The major load should continue until the dial gauge pointer is sharply retarded or the weight arm is no longer controlled by the dashpot, which latter is the better method since it represents a clear final point and one almost coincident with the finish of the crank travel. 2s are now allowed to elapse, after which the crank is *gently* returned to the start. If thrust sharply back it may lead to an error in dial reading.

The maker of the machine provides test blocks for use in checking the machine. These blocks show hardness values corresponding to the metal of which they are made. If the machine reading for these blocks varies from these, it is an indication that some adjustment is required, and instructions for this are usually given when the machine is delivered. Both seat and anvil must be properly seated to prevent vertical motion, or extra depth will be recorded by the gauge and a wrong numerical obtained.

The same point on the testpiece must on no account be struck twice. Should the metal being tested be of thin type or cross-section, the usual precaution of rigid and firm mounting must be taken. In addition the machine must be maintained in a thoroughly clean condition.

Calibrating the Rockwell Hardness Testing Machine The makers of the Rockwell machine suggest that the user should check it each day against the test blocks to ensure that the penetrator is not damaged and the machine not out of calibration, in which case they prefer to be consulted. The practice they advise is to check at the high, low and middle range of the given scale. Thus, to check the full C scale, the machine should be checked at C63, 45 and 25. If only a single range or two ranges are employed, blocks should fall within ± 5 hardness numbers on the C scale or any scale using the diamond penetrator and within ± 10 numbers on the B scale or, indeed, any scale using ball penetrators.

At least 5 tests on the standardised surface of the block are recommended, the average falling within the tolerances shown on the side of the block. The standardised surface only should be employed, as this is the only one having the values marked on the side to which they apply. Curvature error, not always discernible, may occur in the block after a series of tests, so that a pedestal spot anvil is recommended for all calibrations.

If the average of 5 readings falls outside the test block limits, the difference between this and the test block average represents the machine error and must be compensated for when evaluating the results. It is advised, however, that the machine be either cleaned and inspected as instructed in the handbook or the maker be notified.

The ball penetrator or diamond point should always be examined and replaced if defective, as this may be the cause of error rather than any fault in the machine. Indentation spacing is specially important, the distance from centre to centre being at least 3 diameters. If indentations are spaced more closely than this, they should be ignored.

No block whose former impressions have been obliterated by grinding should be employed, as it will probably vary in hardness, while one cannot be assured that the new surface is as hard as the original.

The Monotron Hardness Tester This is a variant of the diamond testing machine, but is little used today. It has two dials (a) for recording the applied load on a diamond ball 0·75mm diameter and (b) the corresponding penetration depth of 0·04572mm. The diamond is given major and minor loads as in the Rockwell machine, the minor load being applied first to penetrate the surface layers. The scale is known as MI.

Other machines using indentation of the testpiece as a measure of hardness include the Elberack, Eberbach, Tukon and Beerbaum micro-character scratch. In all these a standard indentor is used, the load is specified and static, and application is by a system of levers, the testpieces being carried by a rigid platform.

The Shore Scleroscope We have seen that all the indentation hardness testing machines affect the surface of the metal tested, but it may be necessary in some instances that the surface shall not be affected even slightly by the test. In these instances a useful machine is the Shore sclerscope which uses a hammer having a diamond point. This hammer is allowed to fall freely from a previously established height on to the surface of the metal. The operator then measures the height to which the hammer rebounds and translates this into hardness. The hammer weighs about 2·269g (32–40 grains or 0·08oz) and falls inside a glass tube on whose walls is inscribed an arbitrary scale by whose means the rebound is measured. The height of the rebound is registered on a dial gauge, and the fall is 254mm (10in). The dial gauge has 140 divisions.

This test is quick, easy and light, inexpensive and capable of being carried out almost anywhere, since the machine is easily transported. It gives a degree of accuracy sufficient for such parts, for example, rolling mill rolls, that cannot be tested by Brinell or Rockwell machines. The same testing requirements as for Rockwell machines are necessary. The work surface is affected by cold working, so that it is bad practice to strike the same spot twice. Mass and form of the testpiece or part also affect the results, as does case depth in case-carburised work. The usual method is to take several readings and average out the results. The rougher the surface, the lower the readings. The height of the rebound is in effect an indication of the energy expended in deformation of the metal. Tests on light alloys are carried out on occasion at over 200° C (390° F) and BS 1094, 1943 covers the technique.

The Herbert Pendulum Testing Machine This uses a system of weights reposing on a ball of steel measuring 1mm diameter or a spherical diamond, constituting a compound pendulum. The pendulum being placed on the metal surface, the ball just makes contact with and indents this surface, after which it is oscillated through a small pre-established arc. The period of time occupied by the swing measures the hardness of the metal. The load is 4kgf, and in place of the diamond a ruby may be used.

Hardness Conversion Tables Each testing machine is suited to a specific class of work, and in indentation hardness testing in particular, no one property is established, but only a combination of properties. Consequently it is impracticable to convert the hardness values given by one class of machine into those given by another. Nevertheless, tables have been compiled for single classes of metals, such as heat-treated steels, nickel alloys, magnesium alloys, brasses, aluminium alloys, etc. Only those for steel and brass are, however, recognised by the authorities, and are given in BS 860.

Hardness conversions are no more trustworthy than the test results on which they are founded. Every kind of hardness test is liable to errors. Not even the most trustworthy figures can be relied upon to give a single conversion relationship for every metal. The consensus of opinion is, nevertheless, that modulus of elasticity and work-hardening properties are the principal factors affecting conversion relationships.

In making use of Table 1 and 2, therefore, regard should be had to the following points. The conversion values relate only to flat surfaces and to tests carried out in standard machines using all the desirable precautions. The standard steel ball does not give accurate values above 500 Brinell. The Rockwell and other diamond testing machines do not give accurate readings on rough surfaces, but only on very smooth surfaces, and the indentation made gives a good hardness measure only if the testpiece is about 10 × the indentation depth. Conversions from Brinell to Rockwell necessitate tests on metals of uniform hardness to a depth of 10 × the impression. The conversions do not relate to case-carburised or other superficially hardened metals, plated or coated metals or decarburised metals.

Rockwell values are usually to the nearest decimal point, irrespective of the values shown in the tables, and this must be regarded as the limit of their accuracy. Even when it is known that test conditions have been closely controlled, it is found that some variations from the tabular conversion relationships occur.

Ferrous and non-ferrous metals below Brinell 240, ie softer metals, do not lend themselves to accurate conversions, and the labour required in establishing such conversions is rarely justifiable.

Chapter 3

COMPRESSION TESTS

These are far less often employed than tensile tests, as it is commonly supposed that there is little, if any, difference between the strength of metals under tension and that under compression. The tensile test is therefore sometimes taken to meet both needs. This, however, is a fallacy because if reliable design stresses have to be calculated, the strength of the metal in the direction of the service stress *must* be known, consequently compression tests are becoming of increasing importance.

The compression tests are however, less informative than the tensile tests because they have drawbacks. As the compressive load increases, ductile metal testpieces become shorter and increase in diameter indefinitely. Obviously it is not feasible to obtain values corresponding to elongation percentage or to reduction of area percentage, while in addition the tensile test value for ultimate strength has no significance. Another drawback is the greater care required in preparing the testpieces, and the superior operating conditions the machine itself must fulfil. This last factor is, however, no longer so important as it was, quick and systematic compression testing being now achieved without great difficulty owing to the improvement in mechanisms and standards.

Steel in particular supports a greater compressive than tensile stress. The tensile testing machine can be adapted to compression testing as long as it is realised that the action is a compressive rather than a pulling one. The movable crosshead travels in a straight line at right angles to the fixed crosshead, and if this cannot be achieved with the normal machine used, it is possible to employ a subpress introduced between the crossheads. This has a hollow plunger directed by a rigid frame and travels vertically. The load is applied by means of a central rod inside the plunger having enough side play to allow for lateral movement or wobble in the upper crosshead. In this way there is no seizing up of plunger and guide. This load does not make contact with the plunger until it is under the guide, so that there is no increase in plunger diameter and again no seizing up in the guide.

The testpieces have to be machined flat and parallel on their ends to fine tolerances, and must be a minimum of 3 × diameter. In some instances, as with plate material in which the load is at an angle of 90° to the face, the testpieces may be of length equal to diameter. Such short testpieces have higher strength values as shown by the test results than those obtained with longer testpieces.

Compression tests are often carried out on thin sheet metal. This presents considerable difficulty because friction prevents the bearing block, held in contact with the spherical seat, from automatically adjusting to the testpiece face when the load is applied, and in consequence it pivots spasmodically. To overcome this, testpieces with accurately parallel blocks are used, the bearing blocks being adjusted so that their faces, too, are parallel. There are other methods, which space does not allow us to describe. (The Department of Applied Mechanics of Union College, Schenectady, NY can provide detailed information on this point if required.)

The speed of testing has not been standardised, because many compression tests are required for research or inquiry into rather than the acceptance of metals. It is considered by many that the free-running speed of the crosshead should not be greater than 0·0254m/s (0·05in/min), but this would be too high for short testpieces. It is advantageous, however, to employ the same speed for each successive testpiece, making the pauses for dial readings as short and consistent as possible, or creep may occur as soon as the elastic limit of the metal is exceeded, producing an irregular stress-strain curve.

The compressive strengths of materials lacking in toughness, such as cast iron, are usually obtained by the equation $\frac{L}{A} = C$, where L is the maximum load in N and A the initial cross-sectional area. For cast-bronze components, in an American specification (ASTM B22–46T) it is stipulated that one or two proof loads must be obtained for each testpiece, the change of length under each load being measured. This test lays down that for Class A material, the permanent set in 25·4mm under a compressive stress of 689·5MN/m^2 must lie between 1.016 and 3.048mm. For Class B, between 2·54 and 5·08, and for Class C, the maximum compression under 68·95MN/m^2 must not be greater than 0·0254mm.

Other uses of compression tests are for obtaining stress strain curves, either because these give required data or because they are needed for comparison with tensional stress strain curves. High strength aluminium alloys for aircraft, etc, carry their curves up to at least the proof stress, ie the stress at which the offset from the original straight stress strain line corresponds to 0·2% strain as measured by a precision strain gauge.

In all compression tests of metals the strains are measured between gauge points on the testpiece and not between the faces of the bearing blocks. It is considered better that these points should be symmetrical with the middle length of the testpiece and not closer to the testpiece ends than a distance equalling its width.

TESTS OF TORSION

Torsion tests reveal the ability of a metal to withstand the stresses that result in twisting or 'torque'. They are valuable when components such as axles, shafts and twist drills need to be tested to demonstrate their efficiency, and also for materials lacking toughness under twisting stresses. When a component is subjected to twist, each layer of the metal rotates in relation to the other layers and produces a shearing stress. Makers of forgings also use a high temperature version of the test to establish the ability of a metal to be forged, and in addition the ultimate strength of cylinders undergoing internal pressure can be established by use of the torque-twist diagram lying beyond the elastic region.

The name given to the twisting action itself is 'torque', and assuming the forces applied to the extremities of a cylindrical shaft or rod are of the same value and operate in opposite directions, the stress resulting is one of pure shear. This may range from 0 (zero) at the centre of the cross-sectional area of the rod to a maximum at the surface. The strain so produced is also in proportion to the distance from the centre or axis.

The tests are carried out on a machine embodying a lever securely holding one end of the testpiece and giving the required twisting moment or turning effect, by a geared drive, the other end of the testpiece being held in a weighing head which measures the torsional strain given by the first lever. It is essential that the testpiece axis should be coincident with the rotational axis. The measurement of the change in shape of the testpiece is done by telescopes and scales working with mirrors that register the angular deflection of a point close to one end of the testpiece with respect to another point on the same part of the testpiece, close to the other end. The testpiece is usually circular in cross-section and tubular, with thin walls, this form giving a more uniform distribution of stress over the entire cross-section. Care must be taken that the walls are not so thin or lacking in strength that they give way during the test. The testpiece itself has ends larger in diameter than the machined portion that will undergo the test, and there should be no sharp change of cross-section between ends and machined section, a generous radius being employed to prevent failure at this point. The machined portion should also be comparatively short so that the ultimate shear strength can be measured, but if in addition the elastic modulus and the yield strength require to be determined, a somewhat longer section is required.

For ultimate shear strength alone the machined part of the testpiece is $L = \frac{1}{2}D$, L being the length and D the external diameter. Then D should be in a ratio of 10 to 12 of the wall thickness. $L = 10D$ for yield strength and elastic modulus, and the ratio of external diameter to wall thickness should not be more than 10.

Torsion tests have not been standardised, since they are rarely included in purchasing specifications, but the American Society for Testing Materials should be consulted if any specific difficulty is experienced. The equation employed to determine the shear stress is $\frac{16T}{D^3} = S$, where T is the torque and D the diameter. This equation relates only to conditions lying inside the metal's elastic range, so that the maximum shear stress is more exactly given by the equation $S = \dfrac{12T}{D^3}$

More refined figures are obtainable from equations designed to differentiate between tubes and solid bars. Taking S as the shear stress in kgf/mm², P as the force in kgf at the lever arm, p in mm, d_1 the outside diameter in mm, and d_2 the inside diameter in mm, then for tubes the equation is $S = \dfrac{16Ppd_1}{\pi(d_1^4 - d_2^4)}$, but this serves for only those conditions in which strain is proportional to stress. It is, however, extremely useful for determining the values of high stress or the ultimate shear strength.

The complete deformation of the testpiece by torsion is obtained by measuring the angular twist of one extremity of the gauge length in relation to the other. The equation used is $A = \dfrac{T}{L}$ where A is the total angular twist/mm in gauge length, T is the total twist and L the gauge length. A can be converted into shear strain by the equation $S = A(\tfrac{1}{2}D)$, where S is the shear strain in mm/mm and D the testpiece diameter.

To determine the elastic modulus, use is made of the equation $E_3 = \dfrac{SL}{R}$ S being the maximum shear stress, L the gauge length in mm, r the distance from the testpiece axis to the outermost fibre ($\tfrac{1}{2}D$) in mm, and θ the twist angle in radians, in length L.

Torsion tests are used on rope wire to determine its ductility, and the method adopted is to twist the testpiece axially through an angle of 360°. If it fractures before this angle is reached, the wire for rope is rejected. The testpiece is usually 100 × d (diameter). Normally a good wire for rope should give a torsional strength value of 1,234·8 to 1,930·0MN/m² (80 to 125 tons/in²) in the bright surface condition, the precise figure depending on the size. Galvanised rope wire should give the same values, but of course is of smaller diameter: 2·641mm (0·104in) and below to

3·658mm (0·144in) to 2·667mm (0·105in), and the torsion values are usually not so high as for bright wire. The greater the tensile strength of the wire, the lower the torsional strength, as naturally the wire is then less ductile.

IMPACT TESTS

The impact test endeavours to measure the degree to which a metal will withstand the shock produced by impact, and this is done by determining the energy needed to fracture a testpiece when a sudden blow is given. In effect it indicates the toughness, or the absence of brittleness, of the metal. It is impracticable to do more than give here a broad definition of the term 'toughness' as 'the power of a metal to absorb energy and undergo plastic deformation in advance of fracture'. This is an American definition, but it is interesting to note an alternative British definition, namely: 'a condition between brittleness and softness'.

It is sufficient to say that toughness is the reverse of brittleness, so that the greater the hardness and tensile strength of the metal, the lower its toughness. It must be realised, however, that no proportionality ratio exists between the hardness number given by indentation tests and the values obtained from the impact test on an indentical metal. It is true, in fact, that sometimes there is a considerable *difference* between the impact test values of difference testpieces and those of hardness and tensile strength. This is frequently found in rolled metals from which the testpieces have been taken at different angles to the rolling-direction.

For an impact test to have significance, the metal must not bend, but must fracture completely, and to ensure this the practice is to weaken the initial resistance of the metal by cutting a notch in it. The sharper the notch, the more accurate the test results. For example, two different metals gave the results shown below:

	Vee-shaped Notch	*Hemispherical Notch 22mm dia*
Metal A	71·86, 75·93J	93·54, 97·61, 89·48J
Metal B	6·779, 9·491J	40·67, 36·61, 26·93J

The data given by the impact test have to include consideration of the tensile strength of the metal, since high tensile strength alloys cannot be expected to give high impact values. The impact values for a specific metal are frequently more nearly akin to the reduction of area percentage than to any other metal property, and it is usually found that those microstructures consisting of fine grains or crystals give higher values for both impact and reduction of area percent. The fracture produced by the impact test is mostly across the grains, but there are unusual instances in which a highly brittle microconstituent is found along the grain boundaries, producing an intergranular fracture and consequent low impact values.

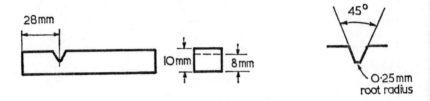

Fig 11

There are various forms of impact testing equipment. In Britain use is mostly made of the Izod machine (1908), but in Europe the Charpy (1905) is preferred. The standard Izod notch is shown in Fig 11. The machine embodies a compound swinging pendulum hammer, released from a predetermined height. The height of the centre of gravity above the level of the vice holding the vertical testpiece being represented by h, the equation used is then $h \times n = E$, where h is the height of the centre of gravity, n is the mass of the pendulum, and E the potential energy, that is, the machine capacity. The pendulum has a standard weight of 27·216kg (60lb) and is allowed to fall 0·3048m (2ft) in centre of gravity, giving an energy before falling of 162·8J (120ft/lb/f) The lower end of the pendulum is provided with a hammer having a hardened metal knife-edge which hits the testpiece at the bottom of the swing at a specific distance above the notch root. The notch itself is set in the same horizontal plane as the top surface of the vice by the adjustment of wedge-shaped dies by hand-wheel, holes in this wheel being introduced

IMPACT TESTS

to accept a tightening-bar. The top of the pendulum ends in a pointer actuating an idle or loose pointer.

The pendulum is first elevated to the specified height, and the two pointers are brought into contact. On release of the pendulum the idle pointer moves to a fixed position in the right-hand quadrant, which has graduations registering the energy absorbed in fracturing the testpiece. The angle of travel beyond the striking position is also registered. In effect the machine records the residual energy of the pendulum, but the quadrant, being graduated in the reverse direction, registers the energy absorbed, since adding these two quantities together equals the machine capacity.

A number of factors have to be added together to produce fracture. First, a crack must be initiated in the metal, and this has then to be extended or 'propagated' across the entire cross-section. Hence the explanation of why close-grained microstructures give comparatively high impact values, since the crack changes direction more often and travels over a longer distance than if the metal had a microstructure of fewer but larger grains. Grain boundaries are stronger than the crystals or grains themselves, so that they are more resistant to the development of cracks.

When rolled metals are notched at an angle of 90° to the angle of rolling, the crack usually follows the lengthened grain boundaries and any fibrous particles of non-metallic material. This involves successive crack regeneration, and the crack covers a longer distance so that the energy absorbed is greater than when the notch lies parallel to the rolling direction. The reason is that in this latter case the grain boundaries or foreign bodies make it easier for the crack to follow a less crooked path. It is therefore advisable to cut the testpieces at right angles to each other.

Moreover, a minimum of three tests should be performed and the results averaged out on each testpiece cut in a certain direction. Hence the testpieces are provided with 3 notches, the second being at right angles to the first and last, which are on opposite sides.

A point of importance is that faulty metals having 'piped'

D

centres, ie centres solidified with a hole running through them from end to end, may nevertheless provide high impact values, for the reasons earlier given. Hence, fracture resulting from impact tests must be studied closely before these values are accepted. The tests show only that a specific metal is tough or brittle at a specific temperature and with a notch of specific form and character. This must be remembered or the test will not be correctly interpreted.

The impact figure is a means of differentiating between various steels in the circumstances of a particular test. The results given are not in complete correlation with shock loading, so that the test does not fully indicate how a metal will behave under impact in service.

The Izod machine is used in the United States as well as in Britain, and having a sharper notch, root radius produces higher stress concentrations than the Charpy, while as it is not so deeply notched, there is greater spread of values over notch-tough and notch-brittle metals. The drawbacks are that the shallow notch is not easy to produce in hard steels, if it can be produced at all, and has limitations when used for tougher steels. Nevertheless, it is widely and successfully used for gun metals, castings, etc. It does apparently discriminate between satisfactory and unsatisfactory metals as regards impact resistance in a way achieved by no alternative mechanical test. Ductile metals may, however, behave like brittle metals on occasion under this test, as is shown by lead, which though highly ductile, gives the values of a brittle metal when impact tested. The explanation seems to be that ductility in the normal sense is given no time to manifest itself when the metal encounters a sharp blow. Moreover the air temperature may affect the results and should be noted.

Welded low-carbon steel plates commonly show brittle fracture, despite having originally great ductility. Again, when tested at extremely low (cryogenic) temperatures, the metal can crack spontaneously, often with grave consequences. The Charpy impact testing machine is often used in inquiring into this problem. The Charpy test uses a horizontal testpiece, the hammer striking it half-way on the side unnotched. The testpiece is $55 \times 10 \times$

10mm, the notch 2mm deep, with 1mm radius. Curves obtained from the results reveal a sharp decline in impact resistance over particular temperature ranges known as 'transition temperature ranges'. These indicate the transformation from ductile to brittle fracture. It is obviously necessary to keep the transition temperature as low as possible, and this has been found to depend largely on the carbon ratio. Rimming or semi-deoxidised steels are much more likely to show brittleness than fully deoxidised steels with a considerably higher ratio of carbon to manganese.

In carrying out these tests certain precautions are essential. The notch must be properly formed and its root carefully located, while the testpiece itself must be correctly inserted in the machine. The dimensional accuracy of the testpiece is normally ± 0.0254mm (0·001in), but if such accuracy is not necessary in particular circumstances, it may be ± 0.0762mm (0·003in) for harder steels and ± 0.127mm (0·005in) on softer materials, with no falsification of the values. (See BS 131: Part 1, 1966.)

The machine must be of standard pattern and properly calibrated, checked for friction and for the zero position of the pendulum. The testpiece must be so placed that the hammer hits it opposite the notch, and a template will enable this to be accomplished. The testpiece must also avoid contact with the sides, which means careful design and maintenance of the machine anvil.

Temperature may greatly affect the values obtained, as indicated earlier, and heat-treatment such as quenching to produce complete transformation to martensite (a hard, needle-like microstructure) and a high tempering temperature often results in a testpiece of steel being of just the right hardness for low-temperature notch toughness. Notched bar impact tests have been employed to indicate the effects of heat-treatment, differences being observed in the Charpy values.

BEND TESTS

Bend tests establish the ductility of a metal under bending stress. They may be simple or intricate. In the simple test, the testpiece is bent mechanically to the desired degree, but without special measurement of the force applied. In the true bend test, however, the testpiece is formed to precise dimensions, supported at the extremities, and a bending stress is applied to the centre, or the piece may be bent about a fixed radius. If the piece fractures before the required degree of bend has been achieved, it is an indication that the metal lacks ductility.

It is often sufficient for the user of a metal to know that it will bend through an angle of 90° or 180° over a former of suitable diameter, without fracture. It is, however, more likely that he will seek to ascertain the precise load necessary to bend the metal until fracture occurs. In these instances the test is not stopped until this point is reached, after which measurements such as elongation, are made as in the tensile test.

Another rough form of bend test is the reverse bend test, in which the testpiece is repeatedly bent backwards and forwards, usually for a predetermined number of times, or until fracture occurs. This is a much more drastic test than the ordinary one, and indicates faults that the normal bend test would not reveal.

Tubes are sometimes subjected to a bend test in which a strip is cut from the tube wall, and a portion of the tube itself squeezed between a pair of plates until its inner walls are a pre-established distance apart.

The bend test of normal type has been standardised in the United States and Britain. The test itself may be regarded as of two distinct types, that for ductile metals and that for hard metals lacking in toughness.

Ductile Metals The old-fashioned bending tests, some of which have been enumerated, have largely given place in testing ductile metals to what is termed the 'free bend' test as long as the metal is not in the form of a complete section of tubing or pipe, etc. For these the test must be between rigid supports and under a bending load.

In ductile metal bend tests the testpieces are usually of rectangular cross-section and have rounded edges. The machine used is a type of press including a roller support and a loading nose terminated by a pin, and can be applied to the testing of either rods, bars, or tubes and pipes. These machines are also used for free bend tests, but it is more usual today to employ a

HARD AND BRITTLE METALS TESTS 53

different type of machine allowing of repeated volume loadings, and also enabling the same bend angle to be developed at every third point. To finish the test a fixture is used, the piece being bent between two compression jaws with diamond knurled faces for the finish bend. These compress the ends simultaneously in a press, vice or special machine. The starting angles are not specially important and range normally from 5° to 30° according to the composition of the metal, but it is more satisfactory if the angles at every third point are as far as possible identical, to make testing easier. The fixture is designed to prevent slip.

Some harder non-ferrous metals and alloys are tested by bending the testpiece around dies of differing radii so that they meet the needs of the test, or to find the least bend diameter acceptable without fracture. For round testpieces the dies have grooves about one third the diameter of the testpiece, with a radius to suit.

The conclusion of the restricted bend test is either that point at which the desired bend angle is obtained with no severance of surface fibres or the point at which the fibres have developed their required elongation. On the other hand, in free bend tests the finishing point is the moment of fracture, and the measurement is taken of the length of the stretched fibres. If fracture does not occur, this same measurement is taken on the flattened testpiece. The rapidity with which the bend is produced is not usually significant except when particular metals are being tested. (See BS 1639, 1950.)

The measurement of fibre elongation is done by laying out the gauge length at the maximum bend section and in the middle third of the testpiece width. The length advised is 25·4mm (1in) for those testpieces above 12·7mm (0·5in) thick. The final gauge length is determined by a flexible scale or the radius of the outer bend elbow may be measured and the elongation calculated from this.

HARD AND BRITTLE METALS TESTS

These include cast iron, hardened tool steels, sintered carbides and on occasion ceramic materials. The test results obtained show a great deal of 'scatter', ie they are spaced out singly with no apparent order, so that the variation may amount to as much as 25% in strength between the same tests carried out on the same metal. It is therefore essential that sufficient tests should be completed to provide a mean value and the range of variation from the mean.

The test is carried out in the normal way, the testpiece being supported at two points and subjected to a concentrated load in the middle of the piece until it breaks. It is termed a 'transverse bend test' and establishes the strength and toughness of the material concerned. The standard 50·8mm (2in) testpiece is measured after fracture, and if it was 50·8mm (2in) long before the test, the values are obtained by the formula: $S = \dfrac{3PL}{2bh^2}$ for a rectangular beam and $S = \dfrac{2\cdot 546PL}{h^3}$ for a round beam. Here S is the maximum tensional stress in the beam section, P is the concentrated load in kg, L is the span in mm, h is the thickness of the beam in mm and b its width. S is, of course, the stress at the external fibres if the testpiece has fractured. There is no fixed correspondence between the bend strength of the metal and its tensile strength in the ordinary way but the bend test on hard and brittle metals does seem to provide a reasonable guide to tensile strength.

The transverse bend test is standard in both Britain and the United States, and covers testpieces round, square and rectangular for cast irons. The testpieces are normally and preferably cast over-size and machined to final dimensions, as this eliminates surface roughness, so minimising stress raisers, shows up discontinuities such as blowholes and cuts out elliptical cross-sections. The dimensions of the testpiece for cast iron should be as near as possible to the thickness of the original casting owing to the varying cooling rates and their influence on the strength of the iron. Square and rectangular cast iron testpieces are usually somewhat smaller than the specifications indicate, as this enables the influence of heat-treatment, microstructure and composition to be perceived.

Bend tests on hardened steels have to be carried out with care, and if so, are more likely than other tests to indicate the fracture strength of the largely brittle steels. In fact, many users claim that they show clearly the ability of a tool steel to withstand breakdown in use. These tests have, however, not been standardised either as regards form and dimensions of testpiece or method of testing, etc.

CREEP TESTS

Creep is the gradual but unceasing deformation of metals under heavy load. A more elaborate definition, is 'the slow stretching of a metal or alloy under stress, particularly at high temperatures, or the continuing permanent or plastic extension of metal stressed at elevated temperatures'. In some metals it takes place at room temperatures, but in others, such as steels, it may occur at temperatures above 300° C (572° F). Steels in general have a wide range of creep properties, some exhibiting great resistance to this phenomenon.

For a considerable period designers of constructions incorporating component parts of metal were unable to determine how they would behave at elevated temperatures. Numerous catastrophes occurred for the simple reason that there were no safety factors or tolerances available to prevent components or structures from failure under these conditions. Such as were adopted did not prevent collapse if the temperature of service was higher than expected. It soon became evident that creep had to be taken seriously and safety factors established for design use. There followed many tests and experiments of long duration at elevated temperatures, and from these in due course a body of data was achieved that enabled true stress values for creep to be obtained.

The creep of metals is usually connected with a rate of deformation continuing when the stresses imposed are much lower than those that will cause the metal to yield, at the specific temperature.

Modern knowledge of creep is largely based on the original work done by the late J. H. S. Dickinson, Chief Metallurgist of the English Steel Corporation, whose researches took a long time for their value to be recognised, but are now regarded with the highest respect. Before his work it was believed that if the service stresses of constructional metals at elevated temperatures were less than those necessary to produce serious creep, all difficulties would be overcome. These values were obtained by normal tensile tests on heated metals, ie short-time high-temperature tests. These were useless, however, because they made no allowance for the time factor. The period of stress duration at high temperature has a notable effect on the properties of a metal, so that the tensile test values proved untrustworthy when applied to those temperatures and stresses at which creep was far from negligible.

The creep test subjects a testpiece to an unchanging load of tensile character at a predetermined temperature, and is prolonged to ensure that a relation is obtained between time and creep or extension at this temperature. Other testpieces are subjected to varying constant loads at this same temperature, and curves are drawn from the data obtained. The highest stress value that can be safely resisted throughout a long service period can then be calculated. No absolute test duration has been established, and tests may last several days or even years, according to the particular metal and the purpose it is to fulfil. Normally, however, a creep test lasts at least 1,000h and even longer in many instances.

How does creep occur? It has been shown in the section on the tensile test that a testpiece extends as it is pulled. When stress is applied to it at high or elevated temperatures there is first a comparatively quick extension, which declines until a virtually constant rate exists. After a time, the rate increases abruptly again, the cross-sectional area of the testpiece notably diminishes and the metal eventually breaks. We thus have five stages: rapid extension, a diminishing rate of creep, a constant creep rate, a new sharp increase in creep rate, and finally fracture.

The practice is to base design calculations not on the final state, but on the most likely efficient service duration of a stressed component at high temperatures, which is taken as corresponding to the period of constant creep rate, or to the creep strain generated during the constant rate period. A sharp increase in creep rate appears to initiate fracture, but not to be solely the result of the load, nor of the deformation of the component, but is caused in the main by heavy modifications of the microstructure resulting from the conditions and period of high temperature service.

The purpose of a creep test is, therefore, to determine the stress at a certain working temperature that is capable of giving metal in a specified time a deformation not greater than a specific rate, eg 1%/1,000h or /10,000h or /500h as may be considered adequate. The time is the estimated service life of the particular component. The load giving this rate of creep at the temperature concerned is called the 'limiting creep strength' at that temperature.

Many of the errors made by designers in the early days arose from faulty measurement of the extension during tests. Today highly sensitive extensometers determine the amount of extension and slight atmospheric variations of temperature affect the instrument's sensitivity and therefore the data obtained. Modern test values are more consistent, but in practice the lower safe stress value is employed if a range is provided.

Creep appears to be influenced by the manufacturing process used for the metal; the heat treatment it may have undergone; the composition and grain size; and the recrystallisation temperature, if any. It is less marked, it is believed, in metals of coarse grain than in those of finer grain, although this has not been fully established. High alloy content is not essential as long as the working temperature is not above 500° C (932° F). Increase in alloy content does not produce a proportional increase in creep resistance. Fully deoxidised steels seem to give better results than incompletely deoxidised steels. Molybdenum, tungsten, nickel, chromium and high manganese seem to improve the creep resistance of pearlitic steels.

Attempts have been made to quicken the rate of testing, and in particular the Hatfield time test was proposed for steels. It gave the stress within which, at a particular temperature, the steel reached dimensional stability at 24h of temperature maintenance, and when heated for a further 48h crept at a rate not above 0·5% of its gauge length during this period. The extensometer developed for this test was so sensitive that a creep of $2·54 \times 10^{-5}$mm (one millionth of an in) could be measured.

Hatfield regarded two-thirds of the 'time-yield' value as the safe working stress, but it was found that the data were trustworthy only above 500° C (932° F). For complete trustworthiness the long-time creep tests are still essential.

A machine for creep testing has been designed by the British National Physical Laboratory, and consists of a cast-iron baseplate supported by three adjustable legs. Four low-carbon steel pillars carried by the baseplate support the straining gear over the testpiece. The testpiece is held in a suitable device of heat- and creep-resistant metal, and has adaptors for use as required. The holders are linked at the top to a straining screw and at the bottom to a loading lever. The testpiece is screwed into the adaptor, which is itself similarly secured to a holder linked with the end of the straining screw. The straining screw is elevated and depressed by gearing in the upper part of the machine, being angularly restricted lest any angular motion should twist the testpiece and force the scale image off the mirrors of the extensometer. Weighing gear is made up of a loading lever and a steelyard set low to make handling easier. A large angular motion of the steelyard is allowed to enable the testpiece to creep a good deal without adjustment.

The load is applied by way of a saddle arrangement at the opposite extremity of the lever, knife edges being used wherever there is a bearing point. The furnace providing the elevated temperature is electrical and so wound that the temperature gradient over the gauge length on no occasion exceeds 5° C. Precise and automatic temperature control are combined with a neutral oxidising furnace atmosphere. A sensitive optical system determines variation in the testpiece length, the sensitivity being of the order of $2 \cdot 54 \times 10^{-5}$mm (1/1,000,000in).

A suitable creep testing machine must embody simple means of applying the load and locating the thermocouple which measures the temperature, and also provide equipment for closely and accurately determining the elongation percentage. The thermocouple has to make firm contact with the testpiece and the testing mechanisms must be as sensitive as possible to prevent the results from being untrustworthy. A series of machines is needed owing to the variable factors.

Chapter 4

FATIGUE TESTS
Fatigue is the decline in mechanical properties encountered in a metal subjected to repeated cycles of stress. Three factors are involved—the range of stress, the mean stress and the number of cycles. Testing is adopted to establish the extent to which a metal will withstand such stresses. (See BS 3518; Part 1, 1962.)

Metals are not homogeneous in microstructure, that is, when observed under a microscope. They are, in fact, crystalline, having cleavage planes and planes of slip running in various directions. If adequate stress is applied, slipping takes place along the boundaries of the individual crystals or grains. The strength of a metal is governed to a considerable extent by the boundary strength. In pure metals or in uniform solid solutions any insupportable stress results in a fracture of the metal along the cleavage planes *across* the grains. When the metal contains brittle non-metallic inclusions, this fracture may occur more regularly *along* the boundary lines.

If the stress range is not great enough to cause slip, little damage may be done, but otherwise the slip, sufficiently often repeated and of adequate size, generates microscopic cracks which gradually extend and ramify until they lessen the area of uncracked metal left. The crack eventually results in a tiny zone of highly localised stress, which enhances the growth of the crack, known as a fatigue crack. Finally the metal fractures without producing any considerable distortion or diminution of the cross-sectional area. Though the fractured surface looks rough and 'crystalline', this has nothing to do with its cause.

The testing of fatigue of recent years has revealed that fatigue strength is normally proportional to the tensile strength, but there are many instances where this does not hold good owing to the wide variations in methods of manufacture, heat-treatment, finishing processes and service conditions, all of which affect the fatigue resistance of a metal. Metals possess what is termed a 'fatigue range', that is, the range of stress to which they are subjected that does not lead to fracture when the stresses alternate a considerable number of times.

The fatigue strength is governed in some degree by the manner of stress application. The stresses may be equally tensional or compressive, or tensional and compressive stresses may alternate successively without any intermission during which the metal rests. If the maximum stress in tension S_{max} is equal to that in compression, the fatigue range is double the fatigue limit or maximum stress in N/m^2 that can be applied without causing fracture within a considerable number of stress alternations. The mean stress, S_m one half of the range, has to be indicated before the fatigue conditions can be defined.

In testing fatigue, failure usually occurs at a much lower value of repeated stress than that of the maximum stress under static tensile stress. Repeated stress tests for fatigue resistance may comprise in the first place a tensile stress, followed by a compressive stress, these stresses being repeated alternately in rapid succession, up to fracture. The stresses applied may be of differing amount, and the test gives the number of stress cycles (S) that can be withstood without fracture. The results are plotted on a diagram (termed the 'S–N diagram') with the maximum stress or applied stress range as the ordinate (vertical line) and the number of cycles (N) required for fracture constituting the abscissa (horizontal line). The proportion of minimum to maximum stress throughout a cycle is stated, and is negative for a cycle of partly or fully reversed stress ($S_m = O$). The test must show the kind of maximum stress, for example whether tensile, compressive or shear.

In relation to ultimate tensile strength, test results for fatigue are fairly consistent for steel, but less so for other metals and alloys. Light metals in particular show wide variation in fatigue strength according to the method of manufacture and working, for example, forging, casting and cold working, while considerable 'scatter' of results is shown by both the brasses and the bronzes.

Much discrepancy can be accounted for by poor machining or

FATIGUE TESTS

faulty design embodying abrupt changes of cross-section, sharp corners, inadequate filleting, and other causes of local stress concentration. Another cause of 'scatter' is corrosion, which produces swift fatigue failure owing to intergranular deterioration. When stress and corrosion both occur in the same metal the result may be catastrophic. Much of this difficulty can be overcome by testing the metal in a carefully controlled neutral atmosphere.

The equipment used for fatigue tests necessarily includes a means of producing cycles of repeated stress; a measuring device to indicate both maximum and minimum stresses undergone throughout a cycle; a counting device to show how many stress cycles the testpiece has undergone; and a means of automatically arresting the machine at the moment of fracture.

There are various proprietary machines on the market, most of which derive their power from an electric motor. On fracture of the testpiece or if there is a considerable rise in deflection or extension, release of a catch breaks the motor circuit and halts the machine. The counting devices are of revolution type.

A British machine gives the testpiece an alternating load developed by an alternating magnetic flux (two phase) working on an armature. The testpiece is placed vertically in the machine, which has a head supported by pillars. The magnets are laminated and of electromagnet type. All the parts are flexibly mounted and the load applied is up to $6 \cdot 227 \times 10^4 N$ (14,000lb) at 180kHz (3,000 cycles/min).

In the United States, however, the most popular fatigue testing machine is the 'R. R. Moore', in which the testpiece is carried in a tapered chuck revolved by electric drive. Two loads are symmetrically applied, and as the testpiece rotates, the weights used produce cycles of fully reversed bending stress. Over the whole length a uniform bending moment is given. The torsional stress resulting from bearing friction is reduced by ball bearings, and the stress is calculated by the normal flexure formula.

An alternative American machine employs a rotating cantilever beam for fatigue test, similar in principle to the 'R. R. Moore' but when used for small testpieces it may be run at 720kHz (12,000 cycles per minute) reversed stress.

Both types of machine are free from unnecessary complications, and are so mounted that vibration is virtually absent. They will measure stresses in cylindrical specimens with great precision, but have the drawback that it is not easy to use them to test the influence of surface conditions, the reduced part of the testpiece being machined below the original surface. The rotating beam machine is also difficult to apply to tests in which there is no full reversal of stress, and if the testpiece is other than cylindrical, speed must be slow owing to the vibration caused when a non-cylindrical testpiece rotates at high speed.

Because of this machines have been developed which are vibratory. In these the testpiece is gripped between jaws that can be vertically reciprocated by a screw producing stress cycles of differing ratios of minimum to maximum stress. For partly or fully reversed stress this ratio is negative. The amount of testpiece deflection governs the complete stress range in a cycle and is controlled by a variable throw crank and connecting rod.

When the location of the jaw deflects the testpiece more in one direction than in the other, there will be a discrepancy between the stress cycle on the underside as compared with the top, even if the complete stress range is identical. Under these conditions fracture normally begins where the tension is at its maximum, and these are the stresses taken into account. Fracture may be initiated on the compression side in asymmetrically cross-sectional testpieces and in those subjected to repeated direct axial stress. The fracture may begin as shear failures.

In ascertaining the arrangements needed to develop a particular stress cycle, the testpiece itself constitutes a dynamometer or energy measuring instrument. Calculating the degree of force necessary to produce the required maximum stress, the operator swings the connecting rod to one side, the load is applied by weights, and the deflection at a particular point along the testpiece length measured by micrometer dial gauge. Enough load is then applied to develop the requisite minimum stress, the reading of the dial follows and the connecting rod is secured to the testpiece. The crank throw to give the required total range of stress is determined by trial, and the jaws are regulated by the screw until the proper ratio of minimum to maximum stress is

FATIGUE TESTS

shown on the dial gauge while the machine is being manually turned over.

If the speed is high, it may be necessary to correct for inertial forces. This means that the extreme dial reading must be recorded during operation, the dial plunger remaining stationary at first, then being adjusted until it makes the least actual contact at one end of its vibration, after which the dial reading and the extreme reading when the machine is manually turned over are compared, the inertial force change being assessed.

This kind of machine is employed when one surface of a metal is in the state characteristic of the work being tested. Thus, it is useful for investigation of the comparative influence of surfaces produced by rolling and polishing on a constructional component. While the machine is in operation the plunger point of the dial gauge is kept from touching the testpiece, and in certain types of machine the specimen is secured to a spring at one end and gripped by jaws at the other, the inertial force being based on the spring deflection.

During fatigue tests the vibrational period may change to some extent and therefore electronic control is used to modify the current frequency, or weights are withdrawn to alter the natural period of vibration. In this way resonance is sustained until the testpiece finally breaks.

The testpieces (usually 6 to 10 in number) are of two different types, and are well-polished, but not buffed. One is intended to eliminate every change of shape liable to increase stress, eg abrupt fillets, notches, grooves, keyways, threaded areas, holes, etc, all productive of concentrated stress in particular localities. The other type produces stress-increasing features. The first class of testpieces reveals the greatest possible strength of fatigue resistance in conditions regarded as favourable, that is, the metal's greatest possible fatigue strength. The second class decides the response of a metal to stress raisers, and when this is high the metals so tested are referred to as 'notch sensitive'. In numerous instances it is impracticable to eliminate stress raisers completely, so that a moderately fatigue-strong metal which is less notch sensitive, may sometimes be more serviceable than a metal of greater fatigue strength extremely notch sensitive.

Surface conditions notably affect test results, and for this reason the polishing of the smaller cross-section should be as axial as possible, that is, lengthwise. Fatigue tests are carried out not only on prepared testpieces, but also on finished parts, assuming these are not too large or unwieldly to be inserted in the testing machine. The machine itself will not give consistent data if its mechanical state is unsatisfactory.

The fatigue limit of a metal varies from one heat to another, and is also influenced by heat treatment for a single level of hardness. The scatter in test results is greater at high than at low hardness.

One testing method is known as the 'staircase' method. In this the testing is carried out close to the fatigue limit, the results being given as mean fatigue limit, that is, about 50% survival. The standard variation obtained from these results is not always trustworthy. A different form of testing is to choose a single heat of, for example, steel, and modify the heat treatment to resemble the variations of production, so establishing how far two heats can have their properties reproduced by the same cycle of heat treatment.

With carbon and low alloy steels tests appear to show that variations in alloy content do not notably influence tensile-fatigue and hardness-fatigue relationships. A carbon content giving a hardness greater than Rockwell C45 also gives a greater fatigue limit. The addition of certain metallic elements as a means of giving greater machinability, such as lead or selenium, appears to lower fatigue strength, so that certain components such as shafts may not be appropriate for a particular function. Sulphur as a facilitating element in machining also has a deleterious effect in reducing fatigue limit, eg in screws. But not only is sulphur harmful when added to improve machinability, it is also detrimental when in the form of non-metallic inclusions in fillets, roots of splined parts, and areas of threading.

It is customary to employ cycles of fully reversed stress in fatigue testing. The equation mostly used is $S_{max} = S_e \frac{(3)}{2-r}$ where r is the proportion of minimum to maximum stress throughout a cycle, r being negative for fully or partly reversed

Page 65 Tensile Testing Machine 7106

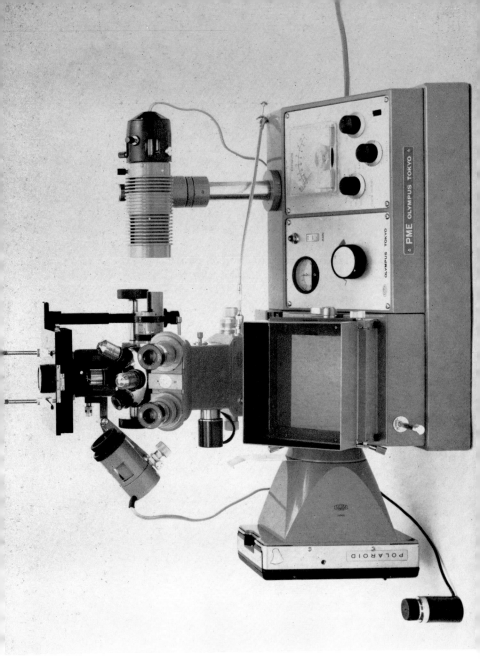

Page 66 The Olympus PME Metallograph

stress; S_{max} being the limit of endurance for any specific number of stress cycles; and S_e being the limit of endurance for an identical number of cycles of fully reversed stress. It is usual to indicate whether the maximum stress is tensile or compressive.

This formula is not so trustworthy as it should be for alloys of aluminium the equation used being then $S_{max} = S_e \left\{ \dfrac{2 \cdot 25}{1 \cdot 25 - r} \right\}$.

For shearing stress in various cycles, an experimental equation is $S^1_{max} = S'_e \left\{ \dfrac{2}{1-r} \right\}$, S'_e being the limit of endurance for shearing stress cycles fully reversed, r being the algebraic ratio of minimum to maximum shear stress, and S^1_{max} the maximum shear stress throughout a cycle.

It must be borne in mind that tests made in the laboratory are merely indicative of how the metal should behave in service, but they cannot forecast the precise stresses to be withstood in service. These have still to be based on past performance of components in actual service, and for this reason it may be essential to test the part or piece under actual service conditions. Nevertheless, if the stress to be resisted in use can be forecast with even reasonable precision, the tests of fatigue strength in relation to dimensions; stress concentration; the state of the metal surface; and the degrees of stress to be encountered; may all aid the designer in preparing details of a component that will prove safe and economical in service.

Today designers decide the effective safety factors by examina- of the stress and cross-sectional area. When aircraft have parts of sheet metal embodied in their construction, failures occur, the primary cause of which is a lack of uniformity in tensional and compressional loading, these being themselves the result of vibrational loading. A great deal of work has still to be done before an effective standard can be attained by which quantitative surface levels of stress and the influence of vibrational stresses on materials can be forecast.

This is particularly true of aluminium and magnesium alloys, whose fatigue fractures show separate cracks on both surfaces of the sheet, which move forward in undulations along two separate fronts, and eventually join up at the point where the axle of zero stress is found, after which they stretch parallel to the direction of rolling. According to Wood and Oxborrow (*Metal Progress*, July 1968) the line along the centre of the fracture surface corresponds to the neutral stress axis between the two crack fronts.

The benefit derived from nitriding steel parts or testpieces cannot be determined from a rotating beam fatigue testpiece when the testpiece is only 7·63mm (0·300in) diameter, as the service components will almost certainly be of larger diameter or size than the testpiece. This is a result of the low tensile stress in the centre of the cross-section directly below the nitrided surface. Consequently no testpiece is able to exemplify the extensive variation in service conditions that may be present in a nitrided steel component, so that all tests should be carried out on the components themselves. This is true also of whatever heat treatment is used to produce fatigue resistance. Every series of conditions develops a particular effect which is by no means easy to forecast from data gathered under widely different conditions.

In the same way it is impracticable to foretell the service qualities of a component in which flow lines result from processes of manufacture, such as forging, especially when these lie at an angle of 90° to applied load stresses.

We have still to deal with fatigue tests at low and elevated temperatures. Very little work was done originally at cryogenic or extremely low temperatures, but tests are growing more frequent in this respect. Not many test results have been published, but the titanium alloys in particular appear to improve in tensile strength with decline of temperature, with a concomitant decline in ductility. Many welded spherical pressure vessels are now made of titanium alloys to form containers for liquid oxygen, hydrogen, fluorine and missile or space vehicle propellants.

The fatigue strength of copper alloys is also important, and at sub-zero temperatures a somewhat improved fatigue strength and toughness is observed, with no loss of ductility, but the rates of creep and resistance to oxidation are comparatively low when compared with those of alloy steels. Consequently they can be safely used at low temperatures for springs, diaphragms, flexible hose, and many other components.

Copper alloy springs have a low fatigue strength as well as low tensile strength, and are tested by load deflection with a testpiece of standard cantilever type, measurements being taken by electronic micrometer. The test results reveal that this metal should only be used for springs where electrical conductivity must be high, the amount of bend small, and the surrounding temperature below 150° C (300° F). Spring temper cartridge brass has a somewhat higher fatigue strength, but load loss takes place at a lower temperature. Phosphor bronze has higher fatigue strength in springs than cartridge brass, but as temperature rises, tests show that it supports a slightly higher load. Little information is available for elevated temperatures.

The testing of stainless steels tempered at varying temperatures can on occasion give values that cannot be correlated with hardness. Some testpieces tempered at 480° C (900° F) show a lower Rockwell hardness than those of the same 'as quenched' hardness tempered at 150° C (300° F). The explanation is that the tempering operation at 480° C (900° F) causes embrittlement, since it corresponds with the convexity in the curve of hardness in relation to tempering temperature.

Notched bar impact testing of the stainless steels also displays a considerable variation in results when compared with those of carbon or low alloy steels. The completely austenitic stainless steels are almost entirely uninfluenced by temperature, the majority of the curves rising, in fact, as the temperatures tend towards the sub-zero range. Notch toughness is almost entirely unaffected by higher temperatures.

THE DILATOMETER TEST

Briefly described, the dilatometer is an instrument for measuring the amount of volume change in a metal caused by modification of temperature or the reversible phenomena (allotropy) in metals showing more than one grain structure.

The early instrument consisted of a large bulb united with a graduated capillary tube and enclosing an inert fluid, into which the solid metal was introduced. When steel is heated or cooled, the volume and also the microstructure were changed, and the instrument determined the transformation temperatures of various metals heated or cooled.

If the coefficient of cubical expansion is known for the metal concerned, the variations/unit length and volume when ascertained enable specific gravity and density to be determined also. These different factors have a relation to one another, and therefore expansion is measurable when temperature varies, or density when the steel's composition is variable.

By means of the dilatometer metallurgists have gleaned a great part of their knowledge of the precise transformation points of steel in relation to both time and temperature. Normally there is great difficulty, even if it is at all possible, in locating with precision the transformation points of steel. The nearest one can get is by lowering the rate at which the heated steel cools to a minimum. On the other hand, by using the dilatometer, changes taking place within narrow limits of temperature can be measured with precision so that the phenomena associated with microstructural change can at once be measured.

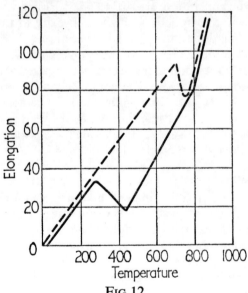

Fig 12

THE DILATOMETER TEST

The record is in the form of a curve, shown in Fig 12 produced by Grenet for a stainless steel. The dotted line indicates the rise of temperature on heating and the corresponding arrest points. The solid line on the other hand represents the fall in temperature on cooling, and again the arrest point is visible. It will be seen that the steel expands regularly up to 700° C (1,590° F) or thereabouts, but then comes a halt or change point when the steel actually shrinks because of the microstructural transformation of alpha iron into gamma iron.

When the temperature is increased to about 820° C (1,380° F) expansion is resumed and goes on with rising temperature. In the cooling curve, however, the steel contracts regularly at the outset, but is arrested in contraction at the temperature of 420° C (785° F).

The value of dilatometric tests is that they provide the metallurgist with an indication of the proper hardening and tempering temperatures for specific steels. Their transformation points being so exactly determined, he is able to heat treat the steels to give the properties within their range that he needs.

There is an alternative method of observing the transformations of metals, namely the continuous—cooling transformation curves, which show the transformations taking place when a metal cools at increasing rates, but this is improved upon by the dilatometer tests because in these the temperatures of change may be either rapidly or gradually passed through, or alternatively the testpiece can be maintained at a specific temperature for whatever period of time is desired, to provide equilibrium, after which heating or cooling may be resumed.

There are many different ways of carrying out dilatometric tests, which can be briefly summarised. The first of these is termed the 'precision micrometric method', which was developed by the American National Bureau of Standards. This can be used in two ways. In the first, by an agitated fluid bath in which the testpiece is immersed in a convenient fluid; in the second, by a heating chamber containing the sample, filled with either air or a neutral gas. The apparatus is suitable for a temperature range from $-150°$ C to $+1,000°$ C ($-238°$ F to $+1,830°$ F).

Whichever method is used, the heating is done by the passage of an electric current through coils, a thermocouple being used for determining temperature and a potentiometer for accurately balancing the potential difference produced against that produced by a current travelling through a resistance. For determination of low temperatures the agitated fluid bath gives the necessary cooling by compressed air in a liquid air interchanger and expands it through coils contained in the bath liquid.

The ends of the testpieces in the air chamber support wires of extremely small diameter 0·0254 or 0·0508mm (0·001 or 0·002in) which hang vertically by means of weights from the testpiece extremities, and pass through openings in the floor of the chamber. The changes in length are shown by the changes in distance between the two wires suspended without restriction from the ends of the testpiece. The wires are normally of tungsten, and the amount of separation between them is measured with a pair of micrometer microscopes set in a horizontal position on a moving instrument. These microscopes are about 100 to 500mm apart. In the agitated fluid bath the wires are slung from a static horizontal rod set over the testpiece, and stretch to hinged fingers placed under the tips of the testpiece. The alterations in length of the testpiece are in proportion to the alterations in the distances between the wires in the same plane as the microscopes.

The testpiece itself is straight and of uniform cross-section, measuring about 300mm long × 5 to 10mm diameter for a rectangular cross-section. The length depends on the ability of the heating chamber to accommodate them, but does not usually exceed 500mm. The microscopes likewise do not approach closer to each other than 100mm when the testpieces are short.

A neutral gas is used in the air heating chamber when the temperature is likely to exceed that at which oxidation or scale forms. Various precautions are taken to prevent vibrations affecting fine measurement, and the compositions of the wires conform to different temperature ranges, for example, chromel, for tests below 700° C (1,292° F); platinum-osmium, platinum-ruthenium or platinum-rhodium for tests from 20 to 1,000° C (68−1,830° F).

On the other hand, the agitated liquid bath may be employed for a temperature range of −150 to +300° C (−302 to +572°

THE DILATOMETER TEST

F). This type of bath gives quicker results than the air chamber because of the agitation and the close contact with the hot or cold fluid. The measurements are recorded as soon as the testpiece is at the same temperature as the furnace, that is the moment the wires become stationary when viewed through the microscope eyepiece. It is essential that the liquid employed should not cause còrrosion or react adversely with the testpiece, and for preference at the normal temperature range a light mineral oil is used, with a proprietary fluid for the sub-zero temperatures. The insertion of a water-filled coil in the bath speeds up cooling between 300 and 20° C (570 and 68° F). It is usual to light up the wires by electric lamps.

The testpieces must be held in a horizontal position throughout and readings must be made at a predetermined distance above each end. The ratio of distances must also be predetermined and maintained.

This is probably the most exact means of determining linear thermal expansion, and given suitable conditions, should be accurate to 0·1 %. Space does not allow further description, but readers should write to the National Bureau of Standards of the USA for whatever additional details they require.

The second method, known as the 'Interference Method', is designed for small testpieces up to but not exceeding 10mm long. It can be used for other investigations also. The testpiece is vertically held between a pair of transparent plates, made of fused quartz and about 4mm thick, without discontinuities such as blowholes or bubbles. The plates are at an angle of about 20min to each other and must be optically flat or nearly so. Both plates and testpiece are inserted in an electric furnace or cooling chamber and illuminated. This produces interference fringes owing to reflected light between the plate surfaces when the angle between them is small. These fringes are studied through a suitable viewer. By modification of the testpiece temperature the length of the testpiece is altered, so that the space separating the plates also changes, resulting in a corresponding movement of the fringes beyond a mark on the bottom surface of the top plate. Alteration in length or linear expansion caused by heat can be detected from the fringe displacement. A thermocouple gives the temperature, or a suitable type of thermometer may be used, either at low temperatures or at those not exceeding 1,000° C (1,830° F).

The linear thermal expansion for a specified temperature can be calculated from the formula: $\Delta L = \frac{\lambda N}{2L}$, ΔL being the alteration in length or linear thermal expansion/unit length; λ is the wave length of monochromatic light; N is the number of fringes beyond the mark, and L the original testpiece length. This formula applies when the testpiece is held in a vacuum during the operation, but if the operation is carried out in air, either for heating or cooling, an 'air correction' is used depending on the pressure and air temperature. The mean thermal expansion coefficient is then calculated by dividing the corrected ΔL by the alteration in temperature.

The third method of dilatometric test is known as the Fused Quartz Tube and Dial Indicator method, the equipment being employed to establish the linear thermal expansion for different ranges of temperature between $-190°$ C and $+1,000°$ C ($-310°$ F and $+1,830°$ F). The test gives a figure of about 2% accuracy.

The specimen or testpiece is inserted in the tube and is about 200mm long. The tube is sealed at the bottom. On the testpiece rests a movable fused quartz rod which stretches over the unsealed end of the tube. Both tube bottom and rod are ground concave, the testpiece ends being made convex to ensure adequate contacts. The upper surface of the movable rod is flat and upon it rests a dial indicator. The tube is inserted in either water or oil, or it may be heated in an electric furnace extending well over the testpiece top. The low temperatures are obtained by some form of cooling liquid. The temperature is read from a thermocouple inserted in the tube close to the centre of the testpiece. The dial indicator records the difference in expansion between testpiece and an equivalent 200mm length of the fused quartz, a slight correction being required for linear expansion of the quartz itself.

This test has been modified in various ways by later workers, but details of these modifications and further information regarding the test can be obtained from the American National Bureau of Standards.

A fourth dilatometric testing method is the Autographic Optical Lever thermal expansion system employed mostly in industrial laboratories. This enables expansion curves to be obtained in the form of photographic records. It can be employed to study particular points in the curves throughout the operation.

The equipment comprises a controllable furnace and a camera box on which it is placed. The source of illumination can also be regulated. The camera holds plates, films, sensitised paper or ground glass screens, according to whether visual or photographic study is to be carried out. The equipment measures 1,626mm (64in) overall, being 482·6mm (19in) tall × 355·5mm (14in) wide.

A mirror maintains contact with three fused quartz rods projecting from the furnace end. A right-angled triangle is formed by lines uniting the points of contact of these rods with the mirror. The 'stationary' axis is the intersection of the two lines forming the right angle which touch the extension at this point. The testpiece is located below this, and expansion produces a vertical deflection of a spot of light by the mirror. This spot traces a curve caused by the vertical and horizontal expansions.

The test gives an error of about 6% for metals which expand similarly to steel through the range 20 to 100° C (68 to 212° F), and about 3% for the temperature range 20 to 500° C (68 to 932° F). The test is not accurate enough for metals whose rate of thermal expansion is low, nor for those temperatures that soften the testpiece or render it subject to flexure at high temperatures. However, within its limits it is reliable, effective and not difficult to use.

The Liquid Micrometer method uses a special form of testpiece consisting of a hollow cylinder 14·05mm ($\frac{3}{4}$in) diameter, the hole being 9·525mm ($\frac{3}{8}$in) diameter. This testpiece is inserted in a horizontal position in a tube of silica with an inside diameter of 25·4mm (1in). The tube is heated by a platinum wire coiled round it to a length of 228·6mm (9in). A pair of silica discs, precision ground, are applied to the testpiece extremities, which have previously been rendered perfectly parallel by grinding and polishing. The discs are held firmly against the testpiece ends by a pair of silica tubes, one of which abuts on a heavy block of cast iron, the other being forced against a measuring device held in position by a large block of lead. These two tubes transmit changes of length to the testpiece.

The tube movement is determined by a steel disc of considerable thickness whose surface is made concave to carry a thin disc of saw steel fastened over it. The space between the two discs contains tinted water. Through the back of the disc runs a tube linking the liquid with a horizontally placed glass tube having a scale attached to it. The quantity of liquid can be controlled by a stopcock from a small reserve quantity.

The liquid measuring device is rigidly secured to the block of lead, the thin steel disc on the other hand being vertically held. A steel disc, 25·4mm (1in) diameter and flat on the surface, transfers the motion of the silica tube to the thin saw steel diaphragm. When the glass tube has an internal diameter of about 2mm, the displacement of the silica tube is magnified about 2,000 times by the curved liquid surface. The testpiece temperature is registered by a thermocouple located at the centre. The joints of the tubes are rendered free from leakage, and means of preventing oxidation are employed. The apparatus is calibrated with testpieces whose thermal expansion is known.

This basic system has since been improved in various small directions.

The Induction Furnace and Dial Indicator method employs a high frequency induction furnace, but its application is solely to heat-resisting materials of ceramic and other types, not to metals.

The Capacitance method is a product of the British National Physical Laboratory, and employs an oscillatory valve circuit. The testpiece, about 20mm long, is caused to produce motion in the moving plate of a capacitor of small size making up a portion of the grid capacitance of the electric circuit. By minute modifications of the capacitance, comparatively significant alterations in the steady anode current of the valve can be produced. A thread recorder continuously registers the current and at the

THE DILATOMETER TEST

same time the testpiece temperature. The length of the testpiece is brought into relation with the temperature.

The Density method does not need extensive description here as its primary function is to establish thermal expansion coefficients of cubical type.

There are other dilatometric tests not indicated, but details of these can be obtained from the national bodies referred to in this section.

The information given by these various tests can be used to show the correspondence between temperature and thermal expansion in the study of the microstructure of metals and alloys; the explanation of volume change in steels during the hardening operation; the study of ageing behaviour of metals and alloys; the kinetics of austenite transformation; the study of the graphitisation of cast iron; and the determination of the relationship between thermal expansion, chemical composition, thermal and mechanical treatment, etc.

Chapter 5

PREPARING TESTPIECES

A testpiece is essentially a portion of metal machined to precise dimensions and made ready for use in a machine for testing. It has to be prepared with great care or the test results will be invalidated. The form and size of the testpiece are largely decided by the type of test, and the specification, if any. Thus, for tensile tests on bars of metal the normal testpiece is made by machining a carefully chosen piece between 15·88 and 28·57mm diameter. About 57·15mm of the length of this piece will be machined down to 14·32mm diameter at the centre, and this reduced length will have a cross-sectional area of 161·29mm^2, rendering calculation easy (see BS 18: 1962).

Testpieces for the different tests have, however, been described in earlier chapters. Here it is necessary to concentrate on the methods of preparation and their justification. The main requirements are that the preparation should be economical, and should not affect the characteristics of the metal to be tested. For example, if the testpiece has been taken from the body of a casting by cutting it out with pneumatic or hand chisel or by an oxy-acetylene flame, certain adjacent zones will have a different microstructure and mechanical properties from the rest, and must therefore not form part of the area on which the test is carried out.

How much metal is to be cut out for use as a testpiece is largely dependent on the dimensions of the main casting, forging or other piece. The preparation of a testpiece from a piece of sheet of relatively thin metal must, however, not be the same as for the testpiece removed from a casting or forging. From these sheets or strips, the testpiece is best sheared to cut away a strip equal in amount to the thickness of the original sheet. This means that for a sheet or plate 6·35mm thick, the testpiece should have 6·35mm sheared from the edge to remove any por-

PREPARING TESTPIECES

tion of metal that might have been affected by the heat of the cutting torch flame.

In the production of some testpieces a punching operation is required to cut out a portion of the sheet material for testing. Such punching has the effect of putting parts of the metal into a harmfully stressed state. If this is known to be the case, it may be necessary to give the testpiece a thermal stress relieving treatment before it is introduced into the testing machine.

Tensile Testpieces The testpieces, usually 3 in number, must be symmetrical about their axes or harmful stresses may be induced in them. They should not involve blanking or shearing operations if this can be prevented, because these may work-harden the edges of the piece, spoil their shape or render them unsatisfactory in surface. The test will then be affected. Nevertheless it is possible to blank out or shear certain thin metal testpieces and by some form of finishing operation, such as filing, shaping, milling or grinding, bring them to the correct dimensions and prevent test discrepancies.

Testpieces to be produced in large numbers may be produced by transverse milling, a required number of the pieces being clamped together in a vice and finished by a milling cutter, but if the quantity required is small, they can be individually finished by various methods or a number of methods combined, such as filing, grinding, shaping, etc. It is important to adhere closely to the specified dimensions and in particular the final machining operation should produce surfaces with smooth and regular surfaces, clean sharp angles and an absence of superficial projections. The extremities of the central machined section should be exact in width.

Large tensile testpieces are finished in milling machines that cut both edges of the reduced cross-section at the same time in a lengthwise direction. The cutters have to be maintained sharp and carefully treated, while the milling operation itself must be carried out with maximum efficiency. The primary requirement is a fully smooth machined surface of the correct dimension and having a microstructure homogeneous throughout and not affected adversely by burning, overheating or cold working.

The testpieces once prepared should be carefully handled and stored until required. Their surfaces must be guarded from notches, indentations, scratches or careless marking. Where the testpiece is of sheet metal, marking is best done with steel scribers whose points are fully sharp, or electric pencils. Alternatively, crayons or inks may be used, steel dies or stencils being avoided.

While the length of the ends of testpieces of sheet or plate metal may be specified, it is probable that better results could be obtained by making these ends somewhat longer. The length given in the specification is in fact a minimum, not a maximum.

Round testpieces are more easily produced, use a smaller amount of metal and provide information of greater consistency than rectangular testpieces. They are usually machined in the lathe, special tooling for large quantities being used in automatic machine tools such as turret lathes or screwing machines. For the light metals special forming tools are used for quick finishing at low cost, the tools being carried on the cross-slide of the lathe.

Where the testpiece metal is difficult to machine, grinding is used for finishing, and for this the wheels must be carefully chosen to ensure that the best form, quality, grit and bond are used under the best cutting conditions of speed, feed and coolant. Otherwise the metal may show grinding cracks caused by overheating. Grinding can also be used for finishing a machined testpiece to bring it to the desired final dimensions. This may be necessary when the steel has been hardened, and distortion has occurred during the quenching operation.

The threads of threaded testpieces must be concentric with the testpiece axis. The holding devices that grip the testpiece ends must give a balanced stress that will not impair the test results. This means that the stress must be uniform and axial.

Tubular products whose tensile properties are to be tested and which are not greater in diameter than 38·1mm are usually tested by inserting metal plugs into them. These are cylindrical for the greater part of their length, but curve towards the entering portion, which is then truncated so that it has a flat surface on the end. No lubricant is used when the plug is forced in, and after the test the plug is extracted, by hammering the plugged

PREPARING TESTPIECES 81

end of the tube to enlarge it. Tubes of smaller diameter can be plugged with an alloy of low melting point instead, the plug being eventually extracted by immersing the plugged end in hot water, the alloy being then retained for further use. The plugs have to be placed carefully in position in the testpiece ends, and to ensure this they are usually machined to somewhat greater tolerances than for ordinary tensile testpieces. The plug enables the testpiece to be held in the machine jaws without collapsing. These jaws do not extend as far as the conical portion of the plug, stopping short where their holding force is applied only to the fully cylindrical portion.

Some types of tubular work are tested in the tensile machine by holding the flattened ends of the tube in flat wedge grips. Ferrous tubing is sometimes tested by means of testpieces of transverse type suitably cut from the product.

Compression Testpieces These have not been standardised as far as the author knows, but three main forms are employed for metals. These tests present certain difficulties because loading, especially in the plastic range, increases in irregularity with increase in the testpiece load. In tensile testing, loading decreases in irregularity when load increases.

The first requirement for a satisfactory test is flatness and parallelism of the testpiece ends, which must also be perpendicular to the axis, with little variation. The finish of the testpieces must be equal to that for the tensile test, and in particular each end of the testpiece and the surfaces it encounters in course of testing must be free from dirt and moisture in advance of assembly. The surfaces concerned are best cleaned by a solvent such as acetone.

The correct type of stress distribution is facilitated by the use of suitable bearing blocks correctly formed and produced. These remedy lack of parallelism of the machine platforms, while other devices may also apply the load axially. When the tests involve testpieces of unusual form or character, other devices again may be necessary.

Shear Testpieces Here, too, there is no recognised standard. In many instances the testpiece is solid or tube-shaped, and is primarily used in tests of torsion. It has thin walls when tube shaped. In one form of the test, the load is measured by a pendulum. The cylindrical testpiece having been placed in position, the torque applied causes the pendulum to rotate until its static moment balances the torque. The pendulum inclination is then transmitted to the spindle of a pointer shown on the machine dial.

The Amsler machine primarily used for shear testing has, however, three annular dies whose axes have square or round holes enabling the testpiece, of indentical form, to be introduced into the three dies. The two external dies are firmly held, the third or inner die being displaced sideways to shear a short piece off the testpiece by means of a fixture in the machine. This constitutes a double shear test giving the shear strength value, but not a pure shear stress, as some of the load is absorbed by friction. A shear box of suitable design minimises bending actions caused by clearances. In this test it is essential that the cutting tools should have smooth, sharp edges, and be well lubricated continuously, the lubricant being applied at the proper speed. The sizes of the testpiece in all forms of shear test have to conform to the dimensions and design of machine and product.

Notched Bar Testpieces In the Izod test the testpiece is prepared by milling three notches 3·301mm (0·13in) in depth with an included angle of 45°, each being 27·94mm (1·1in) from the rest and 27·94mm (1·1in) from the finished end of the testpiece. From these three notches an average is obtained for the value. It is also the practice in certain instances to use a testpiece 10mm square having a transverse notch of 2mm depth and a base radius of 0·25mm. On the other hand, in the Charpy impact testing machine the testpiece is slotted 1mm wide, 5mm deep at the centre, and the bottom is drilled to 1·3mm diameter. The slot is in the form of a keyhole, and the testpiece as a whole is carried by supports 40mm apart.

Since these tests were introduced, much research work has been done and a number of new testing methods have been devised, such as the drop weight test, the explosion bulge test and the tear test, as well as the Lehigh bend test. These are covered in a later section. It has been established that there is a relationship between the temperature at which notch fractures occur and

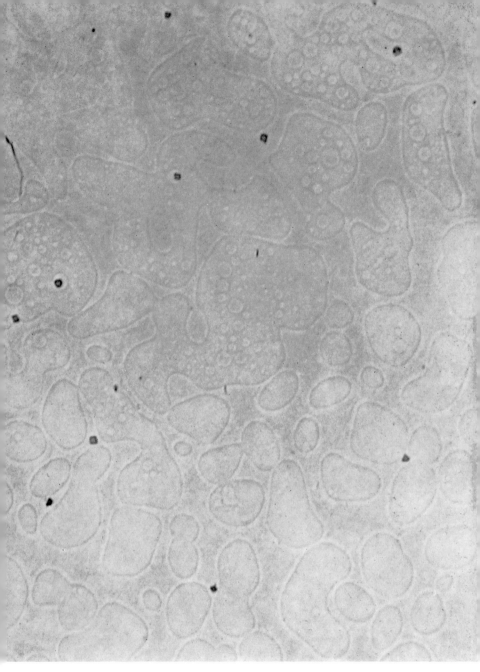

Page 83 Uranium Iron Eutectic 325X under conventional light

Page 84 The same sample of Uranium Iron Eutectic 325X
EPI Nomarski Interference contrast

the transition temperature as determined by the Charpy test, so that this test is usually held to be the better means of testing steel toughness under impact stresses. The vee-notch test is therefore much more often specified by users than it used to be.

While in some closely regulated circumstances the scatter of tests results is slight, it has been found that relatively small variations in the conditions make for greater degrees of scatter, so that no single test should be relied upon. The representative character of the test data obtained from this form of test depends for precision and value on the number of testpieces tested.

The transition temperature has been defined as that marking the transition from a fracture with a bright appearance showing many facets to a silky, fibrous appearance, the test fracture showing the two types of appearance, in equal amounts.

The temperature at which the fracture changes in appearance from fibrous to granular or crystalline is indicative of a change of energy value. Ferritic steels are particularly susceptible to this. The temperature of the transition represents the likely behaviour of the steel under impact. Fractures caused by cleavage take place without warning and quickly, resulting in a breakdown of the component owing to embrittlement.

Another term for the transition temperature is 13·56J (10ft/lb) transition temperature, but this is not a sound definition as it depends on the number of tests performed.

The transition temperature is governed by the chemical and physical properties of the metal, and must be considered in conjunction with the temperature under ordinary working conditions. A component designed for a hot climate may not give the same degree of notch toughness in a cold one if the temperature of the latter lies below the transition temperature. The influence of carbon and the numerous alloying elements also affect this temperature. Those steels of low carbon type produced from ingots completely solid and without occluded gases show the highest transition temperatures, whereas strongly deoxidised steels show the lowest. In between come the steels only partly deoxidised, which undergo less ingot contraction than fully deoxidised steels.

Hot rolling at a low temperature reduces the transition temperature from about +7 to −7° C (45 to 20° F). In a low carbon steel, unhardened, the direction of rolling has some effect on transition temperature, being most effective when the direction is longitudinal rather than transverse, but as the carbon content and hardness of the steel increase, the effect declines.

It has been established that although minor changes in the alloy content of steels, apart from a small number of exceptions, do not greatly alter the energy values of the quenched and tempered materials of largely similar hardness and carbon content, the transition temperature and absorption of energy are considerably influenced by an addition of 0·40% nitrogen. The values of notch toughness or the transition temperatures obtained from the tests on steels quenched and tempered are no guide to the hardness of identical steels tested in a different way.

Case-carburising and nitriding lower the notch toughness of all carbon and alloy steels. The nitrided case in particular renders the surface of a steel harder and less ductile even when the case depth is as thin as 0·127mm (0·005in).

The presence of soft skin, on the other hand, a product of decarburisation of the surface of a steel, is inclined to improve notch toughness, but it reduces fatigue strength and is therefore undesirable, while plating in an electric current also impairs notch toughness, quite apart from the hydrogen embrittlement sometimes encountered as a result of electroplating.

When some steels are maintained at or cooled within a specific range of temperature below the transformation range they are embrittled, as shown by notch toughness tests, and this embrittlement is reflected in the extremely low impact values.

RECENTLY DEVELOPED TESTS

Recently a number of tests not so far described have been introduced, and have enabled certain properties to be determined more satisfactorily.

The End Quench Test This is a method of testing certain steels for their ability to harden, and is applied mostly to medium carbon and the most frequently used alloy steels. The testpiece is a cylinder measuring 101·6 × 25·4mm (4 × 1in) diameter, which is heated to the correct austenitising temperature for the particular steel, and quenched. Quenching involves the vertical holding of the testpiece over an aperture through which a jet of water is forced to play upon the bottom end of the testpiece. The top of the testpiece gradually loses heat to the surrounding air, the rate of decrease of temperature between the two ends varying. The quenched testpiece is ground with a parallel flat on one or each side, about 0·508mm (0·02in) deep, and hardness is then measured at distances of 1·588mm ($\frac{1}{16}$th in) from the fully quenched end. The results indicate the variation in hardness along the testpiece, and the hardnesses are then shown as a graph based on distance from the fully quenched end.

The advantage of this test is that it enables a steel to be bought not on degree of hardness or tensile strength, etc, but on its suitability for certain purposes as determined by its hardenability. It is possible to determine the maximum cross-section at which the material will fully harden to give the specified mechanical properties. For example, it may be necessary for a casting or forging to have a hardness extending about 12·7mm ($\frac{1}{2}$in) deep from the surface, so the steel is bought by reference to its composition and hardenability or 'H band', the graphical indication of the hardness that the steel will give at different points. The graphs are usually accompanied by tables indicating the highest and lowest hardness values at the various distances from the fully-quenched end of the testpiece. It is preferable to select two points to ensure correct hardenability, and the simplest of these is any minimum hardness + any maximum hardness at any desired distance. Alternatively two maximum hardness values at two desired distances may be chosen.

The Explosion Bulge Test According to the American Society for Metals, this relatively new test originated in the brittle failure of plates in welded transport ships. Investigations by the Naval Research Laboratory in Washington DC under the direction of the National Bureau of Standards developed a number of new tests designed to show the notch toughness of the steels with the highest possible accuracy.

In this test the testpiece is 9,033mm^2 (14in^2) × about 25·4mm (1in) thick. A crack is first induced in a weld on the underside of the testpiece, the crack being caused by allowing the force of a carefully controlled explosion to strike the bead, which is about 76·2mm (3in) long. The testpiece is placed above a die, the bead being on the tension side. The testpiece is not bent by the explosion, but is still flat when it fractures, if the temperature of the testpiece is below the zero ductility transition temperature. There will also be fractures of even the die-supported zones on the edges where the explosive force exerted is not so great.

However, immediately above the zero ductility transition temperature there is a good deal of plastic deformation, but the edges show cracks even though the testpiece itself is fractured only with difficulty. Increase of the testing temperature results in the edges being free from cracks, but in the zone at which maximum explosive force is encountered and the testpiece bulges, cracks do appear. The edges show only a degree of shear. As the temperature climbs, so the fractures caused by embrittlement disappear and such fractures as are found result from shear alone (see Fig 13).

The Drop Weight Test Here the method is in the main similar to the Explosion Bulge test. The testpiece is horizontal, and held by partly spherical supports and measures about 82·54mm (3¼in) in length by 19·05mm (¾in). The testpiece is required to bend or

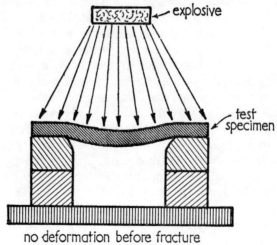

no deformation before fracture
FIG 13

bulge about 5% when struck by a drop weight of 27·22kg (60lb), the metal of the weight being of hard-facing quality. The testpiece has a weld bead which cracks by cleavage at about 3% deflection. A range of testpieces provides the data required to give the zero ductility transition temperature below which a steel having a cleavage crack fractures directly upon the impact without any plastic deformation.

In effect this is the kind of test in which the testpiece either breaks or does not break, the break occurring at the moment of yielding. Its results are readily reproduced (see Fig 14).

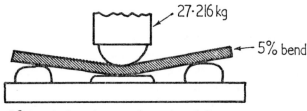

2° bend when drop notch produced. Stop at 5%. Cleavage crack formed

FIG 14

The Notched Slow Bend Test This test discovers the points at which embrittled steel undergoes fracture. The testpieces are notched in a manner resembling the Charpy testpiece, and measure from 241·9 × 50·8mm² ($\frac{3}{8}$ × 2in) long to 5·806 × 609·6mm (9in × 2ft) long. The measurements of load and bend are registered, the notch radius being the same for all tests. The usual tensile testing machine is employed with low rates of strain, the notch being in tension. The notch-toughness of the metal, determined by calculation from the breaking load, measures comparative ductility.

The test is usually applied to metals at room temperature, and it has been found that small testpieces bulge, whereas larger ones fracture, assuming both are of the same metal.

The (Kahn) Navy Tear Test uses a notched testpiece cut out by an oxy-acetylene torch from a complete plate. On the other side from the notch the surface is machined on the edge. There are two pinholes, one on each side of the notch, which accepts pins mounted on shackles. The testpiece is stressed in tensile at a range of temperatures, and measurement is taken of the heaviest load, the energy needed to start and finish the fracture, and the temperature at which tearing begins after the first fracture is discovered. 50% of the shear fracture at a particular temperature represents the transition temperature.

The Lehigh Bend Test This is testing a metal by developing a gradual bend until fracture occurs. The testpiece may be either with or without a longitudinal weld bead, and measures 76·2mm (3in) wide by 304·8mm (1ft) long by 12·7 to 19·05mm ($\frac{1}{2}$ to $\frac{3}{4}$in) thick. The notch is a standard vee or one less severe, and is across the whole width. The contraction width just below the notch is measured, as are the amount of fibrous fracture in per cent and the bend angle at the moment of maximum load. The point at which a sideways contraction of 1% is obtained is plotted against the temperature, and represents the ductility transition temperature.

There are still further slow bend tests, such as the Esso and the Robertson, which differ mainly in testpiece, strain rate, dimensions and type of notch. It is not safe to attempt a correlation of the results of these bend tests as they do not necessarily indicate identical properties.

The Fracture Toughness Test This is a modern test of the strength of steels in thin cross-sections, and has been introduced because of the high levels of stress applied to extremely thin cross-sections, such as the casings of rocket motors, etc, which may need to withstand stresses of 1,379MN/m² (200,000lb/in²). The testpiece is a sheet suitable for tensile testing, and may be either symmetrically notched along the edges or given a transverse slot terminating in fatigue cracks or sharp notches. The tensile test provides data regarding both the load necessary to induce fracture and some aspects of the fractured surface. The usual method of expressing the data obtained is by the equation $S = \dfrac{L}{A}$ where S is the net fracture stress or notch strength, L is the maximum load, and A the net supporting area in the plane of notches or slot.

The strength determined appears to depend on the size of the testpiece and is therefore a variable to some extent, but methods have been put forward to render the results more consistent.

The appearance of the fracture depends largely on the temperature of the metal at the time of test, a wide central band of transverse fracture with chevron markings and narrow inclined shear borders being seen at low temperatures. As the temperature rises, the borders widen until the central band disappears, and this represents 100% shear.

The net fracture stress and the toughness are closely associated with the shear percentage and the behaviour at the transition temperature, but the 50% shear value is considered the safest to use despite the visible 100% shear represented by the fracture appearance. The thickness of the metal is also influential to a considerable degree in fracture toughness, but width and notch depth are less important. Nevertheless, the test is in itself variable in results and cannot be regarded as the final solution of the problem of measuring notch toughness.

Micro-hardness Tests Wear is one of the most serious problems engineers and designers have to meet, and a test devised for measuring the true conditions of hardness in a wear-resistant metal is the microhardness test.

This uses a polished testpiece which has been oil-quenched and tempered to a hardness of Rockwell C30. This is followed by heating in a nitrogenous atmosphere for 48h at 525° C (975° F) to give a shallow, hard case. Hardness traverses are then recorded and often show results not indicated in other ways.

DUCTILITY TESTS

Ductility is the resistance offered by a metal to bending or elongation. In testing for it, the measurement taken is usually of resistance to distortion or 'metal flow' without fracture. It has been indicated earlier that both the elongation and the bend test are used to measure ductility, but there is no general acceptance of these tests as having engineering or design value when applied to this property.

The Cupping Test This is a test of metal ductility in which sheet metals and strip metals are tested for their ability to withstand severe forming operations. The principal cupping tests are the Erichsen, the Olsen, the Amsler, Avery, Guilléry and Persoz. They are extensively applied to metals designed for cold pressing. The testpiece in the form of a section of sheet metal about 1,935·6mm^2 (3in^2) is secured between ring-shaped steel jaws or clamping dies, a ball or steel plunger with a hemispherical end being forced against a face of the testpiece until it fractures, the other face having no support. This results in a cup or depression being formed in the metal, and the depth of the cup at the time of fracture is measured to 0·01mm and normally regarded as an indication of the ductility of the metal, fractures in the direction of rolling, with shallow cups, indicating low metal quality.

All the above-named are modifications of one or other aspect of this test, which, despite its uncertainty, is a common routine test favoured because it gives quick results. The test is so dependant on factors both human and mechanical that the results are extremely erratic. Even the use of hydraulic mechanisms for producing the pressure and automatic pressure maintenance at a fixed ratio to punch pressure have not established these tests as trustworthy indications of ductility.

This being the case alternative systems have been introduced. One of these is the NPL Cupping Test.

NPL Cupping Test Developed by the British National Physical Laboratory, this, using a testing machine actuated by oil pressure, exerts a pressure of 77·25MN/m^2 (5 tons/in^2). The testpiece is rigidly held by clamps between the upper face of the pressure chamber and a die of hardened steel. The head is screwed down to give the desired gripping pressure. Since the head has to be taken out every time the testpiece is replaced, it is in two sections, united by a bayonet fitting. An angular groove cut in the upper face of the pressure chamber contains a leather packing ring upon which the testpiece is pressed down by a ring-shaped tongue projecting from the face of the die. This produces severe pressure on the ring when the head is tightly screwed down. The inside edge of the die has a radius so that it shall not shear the testpiece on application of hydraulic pressure. Depth of cup and oil pressure are automatically recorded.

Other tests on a similar principle are the KWI and the Jovignot Tests.

The KWI Test This is carried out on a machine similar to the Erichsen, and uses a cylindrical tool 40mm diameter having a radiused edge (5mm) and a projection 12mm diameter on the face. The testpiece 90mm^2 is provided with a central hole finished with a reaming tool to a diameter of 12mm. The edges of the hole are carefully finished to prevent roughness. The projection centres the hole at the start of the test, and pressure is applied until a crack is initiated at the hole edge. At this point the hole diameter is recorded and the ductility value expressed as a percentage of the widening of the hole.

The Jovignot Test is closely akin to the NPL.

FORMABILITY TESTS

These are not customarily used in routine testing, but sometimes determine the ability of a sheet metal to be formed by drawing from flat circular blanks. Usually the cups are of a previously fixed diameter formed from a range of blanks whose diameter progressively increases. With each increase in blank diameter there comes a size at which the metal fails when drawn. This diameter is regarded as an evaluation of formability. The machine used is a somewhat modified Erichsen testing machine for cupping tests.

The AEG Test This resembles the cupping tests of Erichsen or Guilléry type, but uses a cylindrical tool with a rounded edge, the testpiece fracturing round the edges of the cylinder. The dies are 50mm diameter. The Guilléry machine is used for the test, since it is claimed to have a more clearly defined end-point than the Erichsen. The testpiece area is usually 90mm^2.

Many other tests have been devised to determine deep drawing or forming quality in a steel, but are not in widespread use.

CALIBRATING TEST APPARATUS

All machines and instruments used in testing metals must be calibrated, that is the amount of their variation from absolute accuracy must be determined. In testing machines calibration implies the determination of loads by subtracting L^1 from L^2, where L^1 is the true load as ascertained by calibration and L^2 is the load recorded by the machine.

Most machines are considered efficient if the permissible error is $\pm 1\%$.

Four types of calibration are recognised by official bodies, namely by standard weights, proving levers, elastical calibration devices and comparison. The loads are normally applied in ascending order and measured in the same way as the machine is used. Ten or more test loads in each range are used in testing calibrations, and some test loads may have to be repeated in case the ability of the machine to provide reproducible results is uncertain.

The load on the testpiece should coincide with the machine axis, and the load range is consequently ascertained by so calibrating that the load is as close to the machine axis as possible. Errors of the machine are found by calibration-loading the machine so that the eventual load lies at specific distances from the machine axis.

There may be no coincidence of error between fast and slow loading rates, the slow rate error often being much greater than the fast, and this point must not be overlooked in determining loading rates for testing. The calibration of test machines must be carried out after all major repairs or modifications of the weighing arrangements or after transfer of the machine to a different location. Otherwise a machine used only occasionally need not be calibrated for 2 or 3 years, but if in regular use, 6 months is the maximum period allowable between calibrations.

Standard weights are available for calibrating those machines of small size whose weighing devices are fixed to the lower platens. These weights may be carried by trays or other supports hanging from the platen, or may be carried by the platen itself. If loading is axial, symmetrical distribution of the weights over the platen is advisable. Both applied and indicated loads are registered for every test load, the error being calculated from the information obtained, but these practices are applicable only to some kinds of vertical machines, and rarely for loads above 917·2kgf (2,000lbf).

CALIBRATING TEST APPARATUS

The Use of Proving Levers These are ordinary levers placed between machine platens in symmetrical pairs, or as a multiple lever steelyard with a ratio of about 10:1. Their capacity seldom goes above 385·9MN/m^2 (25 ton/f) load. They can be standardised and are suitable for calibrating vertical machines. Bellcrank levers are used for horizontal machines and can be standardised. The use of levers is, however, not much favoured as they cannot be used for heavy loads and lessen the portability of the apparatus, so that the majority of testing machines use other methods.

Elastic Calibration This is the most popular method of calibration and in the United States the device favoured is the proving ring originating in the National Bureau of Standards. Usually it is an elastic ring loaded on the diameter, its degree of bend being read from a micrometer screw and a vibrating reed diametrically mounted in the ring. The rings are calibrated by a standard procedure and their capacity range is from 136·1 to 1·361 × 10^{5}kgf (300 to 300,000lbf).

Calibration involves taking readings under precisely known loads. Two types of machine are used, one supporting loads from 90·8 to 4,536 kgf (200 to 10,000lbf), the loads increasing by stages of 45·36kgf (100lbf), the other supporting loads of 907·2 to 5,035kgf (2,000 to 111,000lbf). They may be used for either tension or compression.

An accuracy of 0·1% up to 4.536 × 10^{4}kgf (100,000lbf) is necessary for an alternative testing method developed by the American Society for Testing Materials in which one elastic calibration device is used to calibrate another. The same result may also be achieved by combining a number of elastic calibration devices or by proving levers and dead weights accurate to within 0·0508mm (0·002in).

Elastic device and machine must be at identical temperatures or nearly so. The device must be preloaded within the load capacity of device or machine. Devices called upon to undergo tension or compression may give erroneous values if the stress sets up minor temperature effects, but even greater errors may occur when the device is given a bending stress. Consequently, to prevent these it is advisable to have a fixed time schedule for loading.

Elastic devices should not show an error exceeding $\pm 0.2\%$, and proving rings are expected to have an accuracy of $\pm 0.5\%$ at 10% of capacity load. (See BS 1610: 1964.)

Calibration by Comparison In this method, two sets of testpieces cut from the same bar are tested for tensile strength with both the testing machine and the calibrating process and the results are compared. The method is not so accurate as those previously described.

Torsion Testing Machine Calibration A lever arm whose length is known has forces applied to it which are also known; usually these are dead weights, but an elastic calibration device can be used. The torques applied to the machine weighing mechanism are known.

Hardness Testing Machine Calibration Brinell, Vickers and other hardness testing machines using the method of indentation measurement are calibrated by ascertaining the errors of ball or pyramid form; rapidity and duration of load; and the errors of the mechanisms for measuring the indentation. Not every machine can be directly calibrated, and in these instances calibration is indirect from standard test block hardness readings obtainable from the makers. These blocks are applicable to Vickers, Rockwell and Shore Scleroscope testing machines (see pages 26–41).

Calibration of Impact Testing Machines A simple method of indirect calibration is usual, owing to the difficulties inherent in direct calibration. The product of the total weight of the impacting pendulum and the elevation of the centre of gravity are determined, and scale errors are established by comparing the readings shown with values calculated by the equation $V = W(E)$, where V is the value, W the calibration weight, E the corresponding elevations of the striking edge above the impact position. It is also essential to establish the corrections for friction and windage, as well as the centre of percussion and the striking velocity.

Strainometer Calibration This includes such machines as extensometers, compressometers and strain gauges. The calibration factor here is the ratio of L to P and the corresponding change in machine reading, L being the change in length and P the product of the gauge length. The error is established by a number of different instrument readings, using a calibration device.

A typical device comprises a stout frame free from vibration, spindles of coaxial type to which the machine is secured, a means of moving a single spindle axially with respect to the others, and of measuring the alteration in length so obtained. The measuring apparatus may be an interferometer, a micrometer screw or standard gauge blocks, with an indicator. The errors exposed by calibration should not be more than 20% of the machine's allowable error. There are, of course, other suitable devices.

The calibration factor is expressed as a graph or as an algebraic formula.

Wire strain gauges call for a different method, the chosen gauge being secured to beams or to tensile and compression testpieces, which are loaded to give the required strains.

Chapter 6

SHEET TESTS

While the standard testing machines suitable for the testing of tensile strength, compression, hardness, ductility and resistance to stress and strain are suitable for sheet metals, certain special considerations enter into their use. For example, the results obtained are greatly influenced by the technique and experience of the operator, the accuracy of dimensions and form of the testpiece, the rate at which the tests are carried out, and other factors. This would suggest that the results are variable, which is true, but in many industrial establishments the large number of tests performed without modification of the conditions mean that an average of the data obtained suffices for all practical purposes.

When, however, for research or other laboratory purposes a more exact value is required, as, for example, for proof stress, it is essential to employ a strain gauge of sensitive type capable of determining the 0.2% offset value of this property. In general, the tensile test as outlined in a previous section presents no problems, but elongation is quite often misinterpreted owing to some variation in the degree of taper given to the testpiece from the ends to the central point. Such variations have a marked effect on the elongation percentage and this must be remembered.

Reduction of area is a value not normally obtainable when sheets are tested.

Light cross-sections of sheet or strip testpieces are liable to crumple or distort when tested for compression and should therefore be given lateral support while not producing frictional force. The simplest method is to take a number of testpieces and form them into a pack, but this is an expensive and time-wasting method, since it necessitates machining the upper and lower pack surfaces to ensure that they are parallel. The better method is to press flat pieces of plate against the testpiece surfaces and reduce friction to a minimum by careful control of pressure combined with lubrication. Alternatively lateral support may be given by small rollers.

Only one form of hardness test is regarded as relevant for sheet materials, namely the Rockwell. The load and the type of indenter depend on the composition and thickness of the sheet. The practice of stacking sheets together is a bad one, as it gives untrustworthy values not comparable with those obtained from a single testpiece and a less heavy load.

Ductility tests on sheet metals have largely been covered by comments in the preceding section. The results are influenced by gauge variations and specifically by the width of the testpiece, each decline in width giving a greater cup depth in the cupping test. Bend tests and cup drawing tests have been covered by a previous section.

The test for mechanical anisotropy measures the liability of sheet metals to modifications of mechanical properties when the direction of testing in the plane of the sheet is variable. This produces 'ears' in cylindrical drawn cups. (An anisotropic metal is one whose elastic properties vary in different directions.) The method of evaluating this property is to form a pair of parallel slots in the sheet edge, hold the small tongue so formed between the slots in a firm grip using pliers or other tool, and tear away a triangular piece at a particular angle to the direction of rolling, for example, parallel, 45° or 90°. The length is found to vary with the angle, and shows the anisotropy by its variation. It is also possible to use the ratio of the change of thickness to the change of width, when the metal is tested for tensile strength, as a means of ascertaining the likelihood of ear formation.

TESTING BARS, PLATES AND FORMS

Irrespective of whether these are of ferrous or non-ferrous metals, the tests commonly employed are largely restricted to tensile, hardness and ductility, particularly bend and notch toughness.

Hot-rolled bars undergo more tests than untreated bars, which are mainly tested for tensile or bend values or the two combined. The end-quench hardenability test is suitable for bars of alloy steel with from 0·2 to 0·6% carbon.

Plates and forms are virtually all tested in the as-rolled state by tensile and bend tests, but limited use is made of hardness and low-temperature impact tests, which may be specified for pressure vessels. The testpiece used in such instances is the Charpy keyhole.

TESTING NON-FERROUS METALS

Tensile and bend tests, and to some extent hardness tests, are mainly employed, the hardness tests being Rockwell or Brinell. The properties measured are chiefly yield strength, tensile strength and elongation percentage. The machine used for yield strength embodies an autographic extensometer which produces a curve showing the relation of strain to stress as the test proceeds. Either an autographic extensometer or an indicating extensometer may be employed, but all tensile tests should conform to accepted standards.

The bend test is of value, the testpiece being machined from the metal half-way between centre and external surface of the cross-section, and required to withstand fracture when pressed round a cylinder to an angle of 180°, no sign of fracture being indicated by the bend elbow on its external surface.

Great care is required in testing non-ferrous metals for indentation hardness, especially when the Brinell machine is employed. When the Rockwell machine is used, various combinations of load, penetrator and dials are needed, and the notes on selection of hardness testing methods earlier given should be studied.

TESTING TUBING

This involves the use of most of the standard tests, but some testpieces have to be modified to take account of the curved form and variable wall thickness of the tube. In some instances special tests are necessary according to the purpose of the finished product, the type of metal and its composition.

Tensile tests are commonly used, and for these the testpiece is normally longitudinal and of strip form. The standard tensile testpiece is liable to give a lower elongation percentage than the strip when the tubing has walls of considerable thickness. Other considerations are given in the section on the tensile test.

Some tests are made on short pieces of pipe or tubing by placing them between platens and subjecting them to compressive force, but such a test is restricted in scope to thickness ratios below 10. The test is designed to show the resistance of the tube or pipe to flattening. The pipe walls must not exhibit fractures or cracks within the test limits prescribed. Some non-ferrous metals are also doubled over as well as flattened, the angle of doubling being 90° to the pipe axis.

Bend tests are made as described in an earlier section, and the mandrel around which the metal pipe is bent in the cold state is grooved to take the pipe and is of much greater diameter. The elbow of the bend must not show fracture or cracks, and wherever welds occur, these must remain fully sound.

Tests for Pipe Roundness These are made with a ball or mandrel of predetermined size passed through the pipe. If it does not pass freely, it indicates that the pipe is not of minimum internal diameter.

Tube Expansion Tests In these a tapering pin to specified dimensions is driven into the tube for a stated distance. The expansion must be achieved without fracture or cracking of the tube wall and without any splits in welded portions. The object is to ensure that the tube is sound and the welded zones, usually longitudinal, fully sealed.

Crushing Tests These are intended to reveal the ductility of the tube or pipe metal and to show up superficial blemishes. They are confined mainly to tubes of carbon steel, but some qualities and dimensions of tubes made of non-ferrous metals are also tested in this way. The form the test takes is the application of an axial load to compress the tube until it has lost a predetermined tubular length without fracture or crack.

The test is not of wide application owing to its limitations, and depends for its efficacy on the ratio of external diameter to thickness of wall, which is not easy to determine within a suitable range for crushing tests.

Flange Tests These are simply made to determine if a tube of low carbon steel or a non-ferrous alloy can be safely flanged. A flange formed on one end of the testpiece with a suitable tool is bent about a die block to an angle of 90° without cracking or injury.

Tests for Pressure These may be pneumatic or hydrostatic and are employed to indicate that the tube will not allow liquids or gases to escape. The hydrostatic is the more usual, and two kinds of this test are involved, namely a test for inspection, and a test to destruction. In the proof test a specified brief pressure is applied great enough to establish that welds are sound and joints tight. The duration of the test is not longer than 5s unless there are special reasons for prolonging it. Pressure is restricted to a value not exceeding the formula: $P = \dfrac{2St}{d}$, P being the pressure in N/m^2, S the greatest permissible fibre stress in N/m^2, t the thickness of the tube wall in mm, and d the external diameter in mm.

Bursting Tests These are designed for cylindrical pressure vessels, and, where necessary, tubes and pipes. They are tested to destruction by bursting in the same way as above, except that a hydraulic pump is used to bring the pressure slowly to the necessary amount.

Permanent Set Tests It is sometimes necessary to know how great the permanent set of a tube metal will be when a particular internal pressure is applied. This involves placing the tube or tubular product, such as a gas cylinder, in a water jacket whose water content by volume is already established. On application of the maximum pressure the cylinder in expanding displaces water in the jacket, the volume of this water being then measured. How great a permanent set can be tolerated is governed by the user's requirements, but should not be higher than 5 to 10% of the total expansion for the maximum pressure.

Tests of Collapsibility These are largely applied to pipe which has to withstand severe pressures as in oil well drilling. The pipes must not collapse when fluid pressure is applied externally to the specified degree. Hydraulic pressure is built up slowly and read from a mercury column or a Bourdon pressure gauge, calibrated at regular intervals. The testpiece is 10 to 12 × OD. The wall thickness and the external diameter are carefully measured, before the test begins, to obtain maximum, minimum and average wall thicknesses. The external diameter is also measured at longitudinal intervals.

Collapse of plastic type is governed primarily by the ratio of external diameter to wall thickness (t) and the yield strength of the metal. Values of elastic type are represented by the equation:

$$P = \frac{62 \cdot 6 \times 10^{-6}}{\frac{d}{t}\left(\frac{d}{t} - 1\right)^2}.$$ Here the value of d/t is less than 14.

THE TESTING OF WIRE

There are numerous kinds and forms of wire, some being of ferrous and some of non-ferrous metals, while in addition a considerable tonnage of wire is treated with a protective coating or given a special form of heat treatment, such as patenting. This being so the following notes can only be of general type, and are confined to the mechanical tests.

The main requirement is a knowledge of the strength of a wire and also its ductile behaviour in service. It is therefore tested in the tensile machine to determine the tensile strength, the elongation percentage and the reduction of area. Other tests include the torsional, the bend, fatigue, hardness and wrapping.

The testpieces for these various tests are usually taken from the ends of coiled wire directly after its production or in advance of service, but on occasion testpieces for both are used. The length of the testpiece depends on the user's requirements.

Tensile Tests The ordinary tensile testing machine is used with certain modifications to suit the type of wire, which may be round, flat, or of unusual form. The modifications are largely confined to the use of either hydraulic or screw power, the form of the gripping jaws and their material and treatment.

The results are affected by the rate at which the direct load is applied, which, for a testpiece of 1,524mm (60in) gauge length, should usually not exceed 154MN/m^2 (10tonf/in^2). Proportional limit, elongation, reduction of area and yield strength are other values that may be required, especially for hard-drawn and tempered spring wire. Suitable recording devices are then used as recommended by the machine maker. Great accuracy is necessary for these.

Elongation This value may be measured on the tensile testpiece immediately after fracture, or by extensometer or strain gauge, immediately fracture occurs, by a sight reading. The gauge length should be 254mm (10in) for wire below 3·175mm ($\frac{1}{8}$in) diameter, or for 3·175mm ($\frac{1}{8}$in) and above, 4 × *d*, *d* being the wire diameter. The extensometer readings will give values slightly above those directly measured on the testpiece.

Hard drawn copper wire is measured on gauge lengths of both 254mm (10in) and 1,524mm (60in) according to the wire diameter, the final reading being taken at the moment of fracture.

Reduction of Area The ordinary tensile testpiece is measured by micrometer for external diameter to 0·0254mm (0.001in) for large diameter and 0·00254mm (0·0001in) for fine wire, and after fracture has occurred, the diameter is measured again with a pointed micrometer.

Tests of Torsion These are used for round rope wires, a testpiece being held between jaws 203·2mm (8in) apart, one movable, the other fixed, the testpiece being held straight by a weight fixed to the non-moving jaw. The movable jaw is manually or machine rotated at constant speed and the strength of the material is determined by the counting device recording the number of axial twists through 360° necessary to produce fracture. These are calculated on a gauge length of 100 × wire diameter.

Bend Tests These involve bending the wire 180° to left and right alternatively over profiled edges, the metal having to withstand without fracture a specific number of bends. It is usually found that the greater the number of bends, the higher the reduction of area percentage.

Fatigue Tests These are carried out only when the endurance of wire to be subjected to a large number of alternating stresses is important. The wire surface is not treated in any special way, eg machining, as its initial condition is highly influential. The tests are most often carried out in a rotating strut machine, but if a flat wire has to be tested, a repeated flexion machine is used.

Hardness Tests The Rockwell hardness testing machine is mostly used for flat wire, but if the wire is extremely thin and flat, a Rockwell *superficial hardness* testing machine is better. The limits of this machine are reached when the indenter impression becomes visible on the underside of the testpiece. The ordinary Rockwell hardness test is not confined to flat wires, however, but can be used for round wires also by indenting the round surface and using a correction factor to evaluate the results. The surfaces of wires are usually ground flat if direct readings on the round surface are not desirable.

Wrapping Tests These give a ductility value and are carried out either by hand or with a special appliance. The object is to determine that a metal can be wrapped round and unwound again eight times from a mandrel or former without breaking. They are mostly applied to bronze wire, and iron and steel wires either coated or plain.

TESTS ON FORGED PARTS

These are mostly tested for tensile strength, hardness and ductility, and sometimes for toughness.

Tensile Tests The standard tensile testpiece is used as earlier described, unless conditions require smaller sizes. They are carried out on projections from the forging when this is of large dimensions, but small forgings are sectioned, a sample forging from a batch being cut up. Forgings produced in closed dies are also cut up in this way. The testpiece integral with the forging body or projecting from it has to be deep enough to represent the actual microstructure of the forged piece. The best location is, therefore, the central area of the cross-section or the area most severely stressed.

The ductility shown by transverse tests is usually lower than that shown by longitudinal tests, the explanation being that flow lines produced by forging are more resistant to stress applied across the fibre direction than in line with the direction.

There is a difference in strength between steels whose flow lines or fibre direction run with or transversely to the stresses, while in a heat-treated forging, distortion may occur if the flow lines run the wrong way. It is good practice, therefore, for testpieces to be taken in the direction of maximum distortion, but there are numerous exceptions to this rule, especially where the forgings have to withstand stresses running at right angles to the direction of maximum forging extension in manufacture.

Fewer testpieces will be required for untreated or annealed carbon steel forgings than for heat-treated alloy steel forgings. In some establishments small forged parts are tested by taking a typical one from a batch constituting the product of one melting operation or a single heat treatment furnace charge, but sometimes, as with continuous furnaces, a sample is chosen at intervals as the forgings emerge along the moving furnace conveyor track. On the other hand, the bigger forgings, especially if liable to heavy service stresses, are separately tested, often at more than one point.

Bend Tests The testpieces are mostly taken from that part of the forging representing the upper end of the ingot. The usual method is to bend the metal over a mandrel as exemplified in the notes on bend tests in general. Hard forgings require some modification of the specified bend angle, however. It is not often that more than a single test for bend values is carried out on a forging.

Hardness Tests Mostly these are essential only when no tensile test is made, and are employed to indicate the tensile strength of the metal. The Brinell machine is usually used after sufficient metal has been machined off to remove any soft skin. The point at which the test is made is best settled beforehand by manufacturer and user, as maximum hardness may be required at some particular point or zone.

Impact Tests These are needed only when impact in service is expected. The testpieces are taken in the same way as for the tensile test, but a greater number will be needed to enable an average value to be obtained, as the results vary.

Additional tests sometimes used for research purposes, need not be discussed here.

THE TESTING OF CASTINGS

The principal values required in testing castings are the tensile strength, elongation, yield strength, reduction of area and hardness, but sometimes bend and hardness tests are carried out. The machines used are largely those earlier described, the testpiece being taken from sample castings at particular points

THE TESTING OF CASTINGS

determined in advance in accordance with the service stresses to be encountered and the demands of the specification.

They are usually cast on and project from the piece, being cut off by oxy-acetylene flame or sawn off by circular saw. Another method is to cast the test sections externally to the casting but attached to it, or alternatively to cast special test blocks at the same time as the castings, but apart from it. The advantage of this in manufacture is that there is no interference with the teeming of the molten metal into the mould.

The tests do not so much reveal the mechanical characteristics of the metal as the metallurgical properties.

Bend Tests These are usually called for only when the casting is to be built up into an assembly by welding operations, or when it is likely to encounter shock and vibration in service. The test is a standard one, but the angle of bend may vary.

Hardness Tests These are carried out mostly on the Brinell machine, Rockwell machines being applied only for specific reasons. All the castings may be tested or only a representative sample or samples.

Flexion Tests These are applied to iron castings where it is necessary to ascertain the modulus of rupture or shear. The tensile machine is used, the load being transversely applied to the usual form of testpiece, machined round, its ends supported and its size ranging from 22·22 to 50·8mm (0.875 to 2·0in) diameter by 381 to 685·8mm (15 to 27in) long. The load and flexion leading to rupture are measured and the value arrived at by the equation $E = \dfrac{2 \cdot 546\ SP}{d^2}$, where E is the modulus of rupture, S the distance between the supports in mm, P the breaking load in kgf, and d the diameter of the testpiece in mm. If the modulus of elasticity is also required, this is obtained from the equation $E = \dfrac{P \times 21 \cdot 64}{B}$, where E is the elastic modulus, P the load in kgf, and B the deflection in mm.

Castings are also pressure-tested on occasion, as for hydraulic cylinders, etc, and for creep. Endurance tests are also employed in particular instances.

THE TESTING OF WELDS

Mechanical tests are the lowest in cost and the most trustworthy for weld strength and quality. The tests themselves resemble the normal ones already described, and in general there is little argument concerning either their suitability or the properties they are required to show. On the other hand, considerable difference is found in the form and dimensions of the testpiece and the actual testing methods. Nevertheless, standardisation in these respects is increasing, at all events in relation to commercially welded parts. The American Welding Society and the British Standards Institution have both drawn up recommended standard methods which should be consulted.

Non-destructive tests (see p 112) are important in testing welds, the mechanical tests remaining necessary, however, to detect small cracks, micro-cracks, and some kinds of non-fusion. The testpiece should consist of weld metal alone.

Tensile Tests The usual tensile testing machine is employed and the procedure is as previously described. The testpiece is normally a transverse one, the weld of the joint being in the centre and the properties evaluated being mainly strength and ductility of weld metal and joints. Yield strength may also be ascertained, but is not of great importance. Testpieces are also taken longitudinally.

Hydrostatic Tests These are essentially proof tests to indicate the ability of a welded vessel or other construction to withstand operating fluid pressures and leakage. The fluid used for the test may be water, oil, air or other medium, and the choice among these governs the degree of pressure applied. Usually the pressure in testing is from $1 \cdot 25$ to $2 \times$ the operating pressure specified.

Bend Tests These give a value for ductility if the free bend principle is adopted, in which case the thickness of the testpiece will not affect the results if the width is within the range $1\frac{1}{2}$ to 3 times thickness. For the metal to be considered defective, it must either fail between the gauge marks on the external surface or fracture before being bent double. Cracks on the corners can be ignored, and surface defects must measure at least $1 \cdot 588$mm

($\frac{1}{16}$in) wide or above if failure is to be declared. The best way of determining ductility is by extensometer or direct measurement, but an alternative is to use the equation $E = 100\ t/2\gamma$, E being the ductility, t being thickness and γ the radius to the neutral plane.

Instead of the free bend test, the guided bend test is also used as a pointer to soundness and to the efficiency of the welder. Other forms of bend test serviceable on occasion are the root break and side bend, which, however, have still to be adopted as universal standards. They are intended to reveal unsoundness in weld joints. The root break test is specially useful as a means of detecting imperfect weld metal penetration and unsoundness at the base of the vee. The testpiece is bent backwards to extend the vee bottom in a one-run weld. The side bend test is claimed to show the adequacy of side wall fusion in either vee or groove, and employs a thin testpiece cut transversely from the joint and bent until its weld length is at 90° to the plane of the testpiece.

Drift Tests These are for tubing with thin walls which cannot be tested by the normal testpiece, but which need to have their weld strength indicated. The drift is a steel cylinder tapered for driving into a tube end either manually or mechanically. The mechanical method employing a press gives better results as there is no shock. The drift expands the tube slowly until it is sufficiently stressed to split. This normally occurs externally to the welded zone, and in such instances indicates that the weld is stronger than the heat affected zone and possesses the requisite ductility and strength. The welded tube ductility is measured by the distance over which tube expansion has taken place.

Crushing Tests These are also designed for welded tubes, which are tested for compression under an increasing load until the tube collapses as a result of shear, bending and/or tensile stresses. The primary purpose of the test is to disclose whether or not the tube can be safely flanged, if this cannot be predetermined by the flanging operation itself.

Shear Tests These present no difficulty and allow a welded joint strength to be measured as kg/linear mm of weld failure. The maximum force applied is divided by the sum of the weld lengths ruptured. If the weld is of spot type, two metal strips are lapped and united by a spot weld, after which the test is carried out in the normal tensile testing machine. After fracture, which occurs either through the weld metal shearing or the parent metal tearing, the value is expressed in kg/mm spot weld.

Fracture Test This is an easy and rapid test in which a testpiece is notched on the weld edge, gripped by an anvil or vice, and fractured by a hammer blow. The testpiece measures $1\frac{1}{2}T$ (thickness) in width 12·7mm. The testpiece is not machined, the notch or nick being made by a hacksaw. If the thickness of the piece exceeds 12·77mm ($\frac{1}{2}$in), it is supported by blocks about 152·4mm (6in) from each other, and the blow is applied by a falling tup heavy enough to ensure fracture. The nicks are about 6·35mm ($\frac{1}{4}$in) deep.

This test is not an easy one to pass, and it is probable that any weld failing to break is sound and of good quality, free from porosity, oxide or imperfect fusion. The same test can be used for fillet welds by making an angled weld with two flat pieces and applying the weight or blow to one of the edges, which will open the weld at its root. Any fracture of the weld is visually examined for defects. The failure to break may be the result of too weak an impact, but if other testpieces have broken, under the same force, this is unlikely.

Impact Tests These are carried out exactly as for other tests, but a minimum of 3 testpieces, one notched at the bottom, a second at the side and a third at the top, are averaged out to give the impact in Joules (or ft/lbf) that can be withstood without fracture.

Hardness Tests These are rarely employed except in research work. The one exception is a measurement of the hardness of the butt ends of built-up railway rails.

TESTING MISCELLANEOUS ITEMS
Some small products shipped in packages such as metal cases, casks, boxes, etc, are usually tested by choosing at random an individual piece from the filled container and testing it as a sample of the bulk. Sometimes the test is a rough and ready one, as when a file is used to determine hardness and surface quality. Assuming the right type of file is used, the test does give some

TESTING MISCELLANEOUS ITEMS

notion of whether the piece has a soft skin or is reasonable in its hardness for the purpose of use.

Bolts are sometimes tested to ensure that head and shank are properly united and nuts are tested in a tensile testing machine to see that the threads are not likely to strip off in service. The first of these two tests is carried out by placing the bolt laterally on a steel block, the head slightly projecting over and against the block's square edge. A hammer blow is given to see if the head fractures with a clean break or is plastically deformed.

The thread-stripping test for nuts is carried out by threading the nut on a mandrel of hard steel and applying sufficient load to eliminate the threads. The threads must be fully sound and properly formed if the result is to have value, and the load required is converted into tensile stress, using the mean area of the threaded portion.

Many of these small parts, such as screws, studs, pins, nuts, bolts, etc, are coated as a means of protection against rust, corrosion, and other harmful effects. Normal practice is to measure the coating thickness with an accuracy depending on the purpose. Alternatively the salt spray test is used in which the testpieces are placed in a closed chamber and exposed to a fog composed of droplets of a 20% solution of sodium chloride at a temperature of 32° C (0° F). The test is widely used for electroplated metal coatings and for metals exposed to a marine atmosphere. The testpieces are usually suspended or held so that the solution does not accumulate on them, and it should be a principle that the solution is changed after each batch of tests so that it is not contaminated.

Chapter 7

The tests thus far described have been essentially 'destructive' in that their object is to determine the conditions under which the metal tested fails. This, however, is not the only group of tests employed in industry and research. Many of the modern testing processes do not damage or destroy the testpiece. A considerable number of non-destructive tests are available to the metallurgist and scientific worker, among them radiography, magnetic particle inspection, fluorescent penetrant inspection, magnetic analysis by wave form distortion, inspection by comparing high frequency core losses, electric inspection, supersonic testing, and so on.

Many of these processes require such a highly skilled performance and demand so much detailed technical and metallurgical knowledge that in this section no attempt can be made to give more than a brief outline of each test's function and principles. Readers desiring a full technical account cannot do better than consult the admirable and comprehensive volume on *Non-destructive Tests* (Macdonald & Evans) by J. F. Hinsley, which has earned the praise of many distinguished workers in this field. The tests are of relatively recent origin and play an increasingly important part in testing metals and establishing their microstructure.

RADIOGRAPHY
This is adopted as a means of locating the positions and patterns in which the atoms of a metal are arranged, and of determining whether the metal contains interior flaws such as gas occlusions, hairline cracks, non-metallic inclusions, etc. The radiographic test has been termed the 'all-seeing eye' of metallurgy.

X-rays were first discovered in 1895 by Röntgen, in whose honour they were termed 'Röntgen rays', and are a type of radiation whose motion resembles that of radiant light, heat, radio and other electromagnetic waves. Their wavelength is short, and its variation defines whether the rays are of one kind or another. X-rays are of the order of 10^{-8}cm long. Thus, they are shorter than waves of ordinary light and cannot be seen, but can be reflected, refracted or diffracted like visible light.

Diffraction is the important property for testing. Light waves are diffracted when passed through a prism, breaking up into the separate colour bands of the spectrum. Another method is to pass the rays through a 'grating', a piece of glass ruled with fine parallel lines, and either transparent or reflective. The distance between the pairs of lines is the same in magnitude as the wavelength to be measured. In working with visible light rays several thousands of lines/cm are needed, and as X-rays have a smaller wave-length, it seemed at first impossible to use a grating for their diffraction.

In 1912, however, Von Laue used crystals with regularly-spaced molecules or atoms as diffraction gratings, having concluded that the distances between their molecules corresponded to that of the wavelengths of X-rays. He passed a beam of X-rays through a thin crystal plate and diffracted the rays on to a photographic plate, afterwards developed and fixed.

The pictures thus obtained showed a regular and symmetrical arrangement of dots, from which the exact X-ray wave-length could be mathematically calculated. This length once known, the X-rays could be used to determine how molecules and atoms in crystals such as those of metals could have their spacings established. The photographs also showed the effects of crystal size, internal strain and cold working on metals.

Difficulties of interpretation now occurred, and a thorough knowledge of crystal structure was not possible until Sir William Bragg showed that X-rays would penetrate below the surface of metals and serve as reflecting surfaces or planes by use of the atoms, much as ordinary light is reflected from a plane polished surface, with the difference that the reflected X-rays from the different layers interfered with each other, giving a 'fuzzy' print.

The relation $2d \sin \theta = n\lambda$ had to be fulfilled, d being the distance between two successive layers of atomic planes, θ the glancing angle of the X-ray beam to the surface, λ the X-ray wave-length and n the whole number. This is known as Bragg's Law.

These conditions satisfied, each reflected ray reinforces the others and gives a series of sharp bands on a film so placed as to receive the reflected rays. The bands or lines spaced at definitive distances enable the distances between the atomic planes to be readily calculated, so that the distance between atoms becomes measurable.

By means of X-rays of great penetrative power, interior photographs of metals and alloys are obtained in much the same way as X-ray films of the human interior by medical radiographers. In metals and alloys, blowholes, internal fissures, non-metallic inclusions, for example, allow the rays to pass through them more easily than through the metal itself, and these defects are then shown up by the X-ray film.

In many instances the external surfaces of a metal component are perfect, and only when the component is cut up (destructive) or X-rayed (non-destructive) are the internal flaws revealed. The X-ray test has the advantage that it does not destroy the component, and is specially valuable for the testing of welds. Thus, it constitutes a most valuable tool for the testing of metals. The main limitation is that metal components of more than a few inches in thickness cannot be satisfactorily tested by this method.

This limitation does not apply to gamma rays, which are emitted by radioactive isotopes, substances having identical chemical properties but different atomic weights. Typical examples are iridium 192, caesium 167 and cobalt 60, all of which are used, since all give off the powerful gamma rays. They cost only a fraction of an X-ray set, so that they provide the metallurgist with a comparatively cheap means of radiographic examination.

The essence of a radiographic test is bombardment of the test-piece by a swiftly travelling flow of electrons, which when arrested by a metal have a portion of their kinetic energy transformed into radiant energy or X-rays. The basic requirements are a cathode ray tube as the source of electrons, an anode in the path of the beam, constituting the target, and a difference in

INTERPRETATION 115

potential between cathode and anode to provide the electron velocity as the electrons pass through the metal and give the requisite X-rays.

These requirements are met by means of a gas-filled tube in which the gas is ionised, giving the flow of electrons (other forms of tube are used as well); a highly evacuated glass bulb, cylinder or other form inside which tungsten wire filaments in special tubes give out the cathode rays; a target, which may be solid or hollow; and a set of storage batteries, electric transformers, electrostatic generators and magnetic induction generators, with accessory equipment. The entire apparatus may be fixed or portable, but all concerned in its operation have to be protected, as well as those in the immediate neighbourhood.

A sensitivity of 2% is comparatively simple to obtain and meets most specifications for X-ray tests of castings and welds, and if the various conditions are properly controlled, discontinuities or cavities of 0·5 to 1·0% of the thickness of the metal.

INTERPRETATION
The X-rays and gamma rays give pictures that are in effect shadowgraphs, and the denser parts of the testpiece examined or the entire component show up less than the less dense, which appear as dark areas or bands. Thin cross-sections or deliberately formed cavities such as cored holes in a casting can therefore be readily detected. It is not possible here to indicate every variation in light and shade and its cause, but the following variations are common:
  1. Irregular white or dark zones smoothly formed indicate roughness of surface, differently shadowed, that is, different in intensity, in projections and hollows.
  2. Clearly outlined, globular or round dark zones suggest gasholes, cavities and tiny porosities. (In alloys of aluminium round dark patches freely distributed throughout the entire piece indicate porosity caused by the occlusion of gas, usually hydrogen. Aluminium cast alloys of large grain size show these patches as longitudinal or curved.)
  3. Zones lighter or darker than the test, of irregular form

and varying density, usually signify the presence of non-metallic inclusions.
4. Confluent blotches or spots of both regular and irregular shape and lighter or darker than the rest of the piece usually suggest the presence of segregation, the non-uniform distribution of impurities.
5. Tree-like or hair-like dark areas of varying dimensions and uncertain outline may represent contraction cavities and porous areas of local distribution. If the porosity is not a cavity, it may show a honeycomb or patterned appearance.
6. Dark plumelike lines or uneven dark spots in alloys of magnesium may be caused by microscopic contraction.
7. Highly conspicuous zones of varying size and with clear outlines usually indicate defects in cast parts where the metal has not completely reproduced the mould contours. This is caused by contraction in the crystal boundaries.
8. Dark lines or zones whose lengths may vary and have lens-shaped clear outlines, with one size convex and the other concave, are usually typical of the non-fusion of the metal of a casting where two metallic streams have met, resulting in what seem like cracks or superficial creases or furrows combined with films of oxide. These are termed 'cold shuts' or 'cold laps'.
9. Dark lines whose width fluctuates and whose tree-like branches represent cracks or intergranular tears. Their surfaces have taken on the blue tint found in low carbon steel welds at elevated temperature.
10. Straight dark lines of unbroken type may be cracks in cold metal resulting from too great an internal stress of contractional type, found when the mould is too hard or the casting design imperfect.

Applications of Radiography X-rays and gamma rays are mainly applied to castings and welded components, especially of steel, aluminium or magnesium. In welding they help in the formulation of appropriate production techniques; the routine examination of pressure vessels, boilers, etc; periodical confirmation of the soundness of the less important components; establishment of the quality of zones in which a repair has been carried out.

In castings they aid the establishment of the correct production methods; the routine checking of important components; enable on-site inspection of some or all of the less important castings to be carried out; indicate the quality of welded repairs; and show up production items in intricate castings, among which items may be core nails, chaplets, etc.

In forgings they provide information of value for the operations of production and enable components to be examined as a routine precaution, although this is not an important application, since forging defects are not readily discernible by X-rays or gamma rays.

More important is the use of the radiograph in checking constructions, when it enables the position and size of interior parts to be seen; checks the properties of soldered and brazed assemblies; and indicates the presence of foreign bodies in intricate constructions, such as discarded or carelessly mislaid tools.

Gamma Rays These can be produced by radium, radon (a radioactive gas) or mesothorium (a radium isotope), and give higher sensitivity than X-rays owing to their low absorption and scattering ability, which enables clearer pictures to be obtained. Nevertheless, they are less sensitive than X-rays and have a factor of from 2 to 5%. They are given off spherically, so that a wide range of testpieces can be spaced around the source and studied at one and the same time. Nevertheless the test is not cheap and takes up a great deal of time, while the number of exposures is comparatively small in comparison with X-rays, but the method is advantageous where simplicity of testpiece exposure and the low cost of the equipment are advantageous.

ELECTRON DIFFRACTION

In 1927 it was shown that beams of electrons could be diffracted by a crystal, and indeed, when a beam of electrons was aimed at an extremely thin testpiece having a photographic film behind it, the beam making a small angle with the surface to be examined and the plate receiving the electrons that 'bounce off' or are reflected, a new tool was available to metallurgists.

H

This in no way supplanted X-ray diffraction, but supplemented it. Electronic diffraction takes place when the lattice structure of a crystalline substance has a periodic *potential* variation associated with it, so forming an 'electric grating' that diffracts electron beams.

The wave-length of these beams is shorter than that of the X-rays used for diffraction, which, when generated at about 25,000 to 75,000 volts range from 0.50Å to 0·17Å (1 ångström unit, Å, equals 10^{-8}cm). In contrast, electron beam wavelengths are of the order of 0·05Å. In consequence θ can be used instead of sin θ in the Bragg formula earlier quoted.

The work is done in vacuum because the range of the electron beam in air is short. The high power of matter in arresting the passage of electrons means that testpieces must be exceptionally thin, of the order of 0·5 millionth of a centimetre.

By this diffraction thin surface layers of a metal can be studied owing to a reflection of 'grazing incidence' technique. A beam of electrons just grazes the surface, whereas X-ray beams penetrate into the metal.

The testpieces for this technique have to be prepared much more carefully than for normal X-ray examination. Fingerprints or films of oxide may invalidate the results by hiding the underlying material to be examined. Adsorbed gases in testpieces have to be avoided. The atmosphere of the room in which the test is carried out must be pure. An unseen film of oily or greasy matter thick enough to invalidate the results can be deposited on a testpiece in less than 30 minutes.

This extremely sensitive surface test includes studies of the Bielby layer formed in polishing, the protective surface layers that form on stainless steels, the 'pick-up' on bearing surfaces, and electroplating and wire-drawing processes.

ELECTRON MICROSCOPY

The wave-like attributes of electrons make it feasible to focus an electronic beam with magnetic or electrostatic lenses, in contrast to the glass lenses used for focussing light beams in the normal microscope. Magnetic focussing of light is not uncommon, and virtually all television tubes employ the principle. The property possessed by electronic beams of reflection from the surfaces they encounter renders possible a practical electronic microscope using electronic rather than light beams.

Although these microscopes are costly, the short wave-length of light they use increases enormously the limit of resolution, that is, the size to which the smallest distinguishable feature can be magnified in an object. Normal light covers wave-lengths of 4,000 to 7,000Å, whereby objects less than 5,000Å cannot be distinguished with the normal microscope. The wave-lengths of electronic microscope beams measure about 0·05Å, and therefore the resolving power of an electron microscope is higher than 10Å so that direct magnifications of 100,000 or more have been achieved. Taken at such magnifications photographs can be enlarged about ten times, so that a final magnification of 1 million is now regularly attained with good definition, large molecules of matter being brought into range.

At the other end of the scale the electron microscope can be used down to 200 magnifications or so, allowing a good overlap inside the normal light microscope range, whereby features revealed by both techniques can be identified and correlated.

There are important differences between the techniques of the two forms of microscopy. In the metallurgical microscope the light reflected from the polished metallic surface under examination is turned through 360° to reach the objective lens, but electrons are difficult to 'bounce back' in this way, most reflection work being carried out at low angles of 4° or thereabouts. Consequently incidence and reflection in electron microscopes produce foreshortened and distorted micrographs.

As an alternative a replica method is used, involving the extremely thin coating of a homogeneous plastic substance such as nitrocellulose on the surface of the objective. The structure of the plastic limits the magnification, and details smaller than about 200Å may be lost. The carbon replica method, however, allows observation of details down to about 10Å and has interesting potentialities in the study of metal surfaces by the extraction replica technique.

The limitations of the electron microscope are that the interior and the object studied must be under high vacuum, while the electron beam carries so much energy that it can easily burn through a replica, and can also deposit contaminating matter as a film over the object at about 1Å thickness per second, obscuring some detail. Despite this, the microscope has helped in some valuable work on the microstructure of metals, having revealed the microstructures of bainite (a microstructural constituent) in heat-treated steels, for example, while the study of stainless and heat-resisting steels and the verification of forms of precipitation never before seen, though theoretically suggested, have proved effective.

In using the electron microscope the illuminating system has to be correctly adjusted; the aperture diaphragm must be the best possible; and the wave-length of light employed as well as the means of lighting the testpiece must be chosen only after careful consideration. No set rules can be laid down for adjustment of the illuminating system, the designs of the various microscopes varying so greatly. There is a best possible opening for every objective, with internal reflections or 'flare' so small as to cause little trouble. A small field of maximum resolution is superior to a large flat field with poor resolution. Anti-reflection coatings enable the aperture to be increased to the maximum possible without flare.

The shortest possible wave-length should be chosen and the objective properly corrected to suit, yellow-green for achromatic and blue-violet for apochromatic objectives. (In an achromatic objective, aberration of the type caused by the difference in the index of refraction for light of different wave-lengths, prevents a plain lens from focussing at a single point light from the same source, whose wave-lengths vary. An apochromatic objective is one whose spherical and chromatic objectives have all been corrected as fully as possible.)

Light filters are employed in ferrous metallurgy when choosing a wave-length, owing to the absence of colours in most ferrous objectives; but they are also used for non-ferrous metals, whose colours are thereby more precisely reproduced in tone. If the emulsion employed varies in sensitivity to the different colours, a filter for suppression of those wave-lengths to which it is most sensitive is useful.

Dark field illumination is superior in certain respects to normal bright field illumination. Dark field illumination is achieved by a device embodied in some microscopes of inverted type, and enables the objective to be illuminated from every side at an angle other than one of 90°. It gives less flare and full aperture is possible. High contrast of black and white can be achieved. Less highly polished surfaces are required. Detail not visible by bright field illumination is brought out, and tinted microstructures are revealed in their natural hues. When dark field illumination is used it is often conical in form, as this, when correctly employed, renders the ensuing micrographs uniform in illumination and of attractive appearance, while there is no difficulty in seeing even fine detail.

For the best results from the electronic micrograph the surface of the objective must be polished to show up the real microstructure, and this involves etching to a depth that leaves the detail unmarred and capable of photographic reproduction, while remaining contrast. Maximum resolution must be achieved by careful microscope adjustment, and the exposure should give an image not exceeding the limits of the film or plate and developed to a contrast suitable to the paper used in printing; the paper should have the longest possible contrast range.

Proper exposure involves preventing the brightest areas from over-exposure and the darker from under-exposure. The accuracy of exposure must be greater as the contrast obtained by etching increases. Etching with too great a contrast is not advised. Developing practice should be standardised, while fixing and washing are also to be carried out with care. The bath temperature should be below 23° C (75° F). A hardener should be added to the bath to prevent softening of the emulsion. Fixing is done by leaving the plate in the bath for 3 to 5min with no light on first, then fixing for double the period of time necessary to clear it, and washing with running water to 30 to 60min.

The paper used for printing should suit the contrast of the negative and must show the maximum detail of its darkest area. Negatives of high contrast should have papers giving low contrast and flat or soft negatives should be printed on hard papers giving maximum contrast. Most papers have a shiny surface to ensure this.

ULTRASONIC TESTING

The sound frequency range termed 'ultrasonic' is now widely employed in industry for many different purposes, but for testing its greatest value lies in its ability to detect hidden flaws in metals. This means concentrating a beam of sound vibrations upon a testpiece and studying the reflections of the beam given off by the flaws. The word 'ultrasonic' means that the waves or vibrations lie above the range of audible sound as far as the human ear is concerned, since they are pitched too high. Nevertheless they constitute a form of mechanical energy since they represent vibrations of the air, but should not be confused with electromagnetic waves or rays.

As the pitch or frequency of a sound increases, its wavelength diminishes. The wave-length in air of A above middle C is about 0·6m, whereas the A at the top of the pianoforte keyboard being 3 octaves higher has only a 9·52cm wave-length. A beam of sound waves to find concealed defects must be narrow, so that the device generating the sound must be of diameter several times greater than the wave-length of the sound to be radiated.

In normal sound wave frequencies the generator would inevitably be too large for use, and also not sensitive enough to small defects. Extremely high frequency waves (ultrasonic) are therefore transmitted into the object by a transducer, which converts electrical energy into mechanical motion and vice versa. The ultrasonic waves reflected by the defects and other features of the object are likewise picked up by the transducer and amplified to actuate a cathode ray tube or other detection device, enabling visible signals corresponding to the location and size of the defects to be indicated.

Fig 15 shows the principal characteristics of an ultrasonic flaw-detection device. The internal beam rules on a screen a horizontal straight line upon which echoes are shown as 'blips' or peaks. Any 'trace' or line built up on the screen is in actuality the successive positions of a single light spot travelling at high speed. The human eye retains for about 1/20th of a second the

ULTRASONIC TESTING

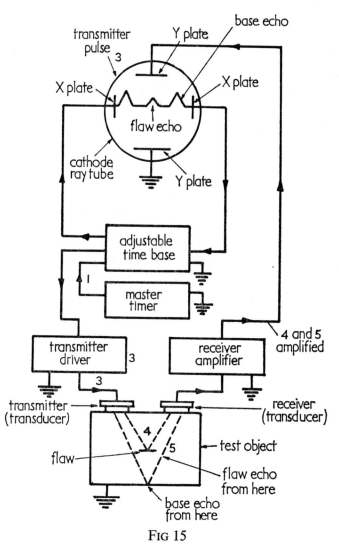

FIG 15

impression of anything seen, and as the trace is repeated faster than about 20 times/s, it appears as a continuous line in the same way as a cinematograph picture is made up of a series of rapidly following photographs.

The master timer indicated in the centre of the diagram emits an electrical pulse marked 1 at a frequency ranging from every 200th of a second. The timer is 'fixed' since it is pre-set and needs no adjustment, but in some sets master timing is done by the mains, and the pulses are only every 5th of a second.

Pulse 1 is transmitted to the component marked 'adjustable time base', which can allow for variations of metal thickness, and it activates the time base to set the light in motion across the screen. The path or trace of the spot is shown diagrammatically between the two X-ray plates. The time base itself, by the pulse marked 2, activates the transmitting transducer, normally a quartz crystal or similar device. Pulse 3 is entirely different from 2 and 1, being a brief 'burst' of electrical vibrations instead of one short voltage pulse. These vibrations are converted into mechanical ultrasonic energy or waves by the transducer.

The transducer or 'probe', as it is sometimes termed, is linked up with the object by a thin film of oil or other liquid and transmits the waves into the object in a fairly narrow beam, as shown in Fig 15, and the waves are reflected. Two reflections are indicated, 4 being from a flaw, and 5 the base echo. 4 traverses a shorter path than 5 and gets to the receiver first. The receiver is also linked up with the object by a liquid and turns both echoes into electrical impulses again. These are fed into the receiver amplifier and highly magnified in advance of passing to the Y-plate of the cathode ray tube.

Every 200th of a second or so a spot of light begins its journey from left to right and presents to the human eye a straight line on the screen of the tube. As its journey starts, the transmitter is set in motion and a small fraction of pulse 3 leaks (see dotted line) to the receiver and so on to the tube. The light spot goes back to its normal position and path until it is again deflected by the arrival of 4 and 5. Obviously, the nearer the defect to the upper surface, the nearer blip 4 will be to the left-hand end of the trace. Skill and experience are essential for interpreting the screen traces, and reference specimens or standards should be kept on hand for control and checking. A big defect can wholly smother the base echo, for instance, and any high degree of porosity can stop all transmission.

Frequencies range from 0·5 megacycle (500,000 vibrations/s) to 20 megacycles/s. At the usual frequency of 2 megacycles/s, the wave-length in steel has a velocity of 5,116m/s (16,586ft/s), and defects of this order are readily shown up as peaks on the screen. Large components such as aircraft wing spars are fully submerged in a water tank, the transducer scanning the spar and registering defects on a chart. The transducers employed in manual scanning are adapted to excite varying types of wave-lengths, which may be either longitudinal, shear or transverse, or surface waves, depending on the character and orientation of the expected defects. Sometimes two separate probes are employed, or two transducers may be embodied in the one probe. Again, one transducer may be used to do the work of both transmission and reception.

The control of internal quality in steels has been much improved by the ultrasonic tests, and has been widely used for the higher-priced steels such as tool steels, especially high speed steels, die casting dies, extrusion tool blanks, die blocks, cutters of high grade steels, etc. In this way the presence of defects can be detected before any costly machining work has been done.

MAGNETIC CRACK DETECTION

Visual and other inspection of metal parts is not always effective. A crack, for example, may be so fine that even if the metal is struck with a hammer it still rings true. Magnetic crack detectors have been found highly advantageous for exposing these minute flaws.

The principle is that after cleaning the metal, usually steel or iron, and eliminating dirt and grease, the work is placed across or between the arms or poles of the machine. These are adjustable to receive parts of widely varying forms and dimensions. An electric magnetising current is then passed through the work or piece to be tested, which is then taken from the machine and a special ferrous powder or solution scattered over its surfaces. This powder is provided by the machine supplier. Any excess is blown or shaken off, and even the finest cracks are shown up as an 'indication', that is, a pattern of powder distribution representing the discontinuity on the surface.

For example, cracks in a ground surface may appear as a 'craze' or network of fine cracks, and are often the result of excessive grinding of a hard metal. Though they do not usually penetrate far below the metal surface, they frequently act as initiators of later deep cracks when the part is put into service. The magnetic crack detector is quick and non-destructive, so that it has many advantages. The machine is easily portable and may be linked up with an AC or DC supply. It is of great service in testing finished components. The surface of the piece is not harmed, and the finest discontinuities are clearly indicated. Many components can be tested while still only semi-finished and the faulty ones thrown out, so eliminating costly finishing operations on worthless metal. Raw materials can be tested for surface cracks or a section machined or ground off can be tested for internal soundness.

The most satisfactory indications are given by machined or ground surfaces and scaled components after heat-treatment, all of which can be locally cleaned at a point of suspected flaw and magnetically tested. The test is mostly applied to billets, bars, tubes, forgings, relatively thick sheets and oddly-formed components such as parts of machines and constructional items. Non-magnetic metals cannot be tested in this way, however.

It is worthy of note that in addition to surface defects the test will sometimes show up defects lying just under the surface, such as large gasholes, large cracks, and so on. The smaller the angle of the magnetic flux direction, the less the strength of the indication. Magnetising should normally be carried out in two directions with the field of one at right angles to the other. For large gashole determination, however, a single direction is preferable, and a hole about $2\frac{1}{2}$cm below the surface, can be detected, if large enough.

The form and dimensions of the object to be tested influence the field developed by a specific current strength, and so do the magnetic properties of the ferrous material itself. Large pieces such as castings or forgings have the current passed through particular zones by means of clamps or forms of contact fastened to flexible cables so that the current can be directed as required. Almost any field direction can be given in this manner.

The powder or 'dry' method uses a fine powder applied by atomiser or spray gun, though simpler apparatus may be employed if desired. The powders can be obtained in colours to give the type of indication most suitable for the material.

The 'wet' or liquid method employs a suspension of magnetic particles in a light oil, carbon tetrachloride, or similar medium with a flashpoint of at least 60° C (140° F) and a maximum kinematic viscosity of 3 centistokes at 32° C (100° F). The liquid may be directed in a jet from a hose over the component or the component may simply be placed in the tank. For this purpose the ferrous particles are usually provided as a paste of black and red tinge, the black used for ordinary surfaces, the red for dark surfaces. About 6g/l (1oz/gall) is the usual concentration.

Protective coatings on the metals do not prevent effective indications. Cracks are revealed as a clear and distinct pattern and can be detected by low field strength, using the residual method in which the medium is not applied until the current has been discontinued. (In the continuous method the current is left on, so that the field strength is directly provided by the metal's magnetic permeability and the current.)

Seams show up as straight, clear, thin and sometimes broken lines, and for their detection a higher field strength is required than for cracks. Inclusions of non-metallic type lying immediately below the surface are exhibited as wider discontinuities with less clear boundaries, and for their exposure require a high magnetising field and the continuous technique.

It is advisable to demagnetise any ferrous or magnetic metal to be tested to prevent foreign bodies from being attracted, such as iron filings or turnings from machine tools. The apparatus may be static or portable.

FLUORESCENT TESTS

Austenitic and non-ferrous metals are not suitable for magnetic particle tests, being non-magnetic, and the same applies to manganese steels of austenitic type (11–14% manganese). Instead, the fluorescent penetration method is used, involving certain substances that fluoresce or give off a form of light known as 'black light' produced by the action upon them of ultraviolet light. This emission of visible light is the result of ultraviolet ray absorption by the metal and its conversion into light rays.

The fluorescent substance is soluble in a penetrant or liquid of light character such as an oil, and when applied to a metallic surface travels into fine cracks and other flaws of the surface invisible to the naked eye. After the applied penetrant is wiped off or otherwise removed, the object is studied by means of an ultraviolet lamp in a darkened chamber. The penetrant that seeps into the discontinuities glows brightly under the lamp rays and shows up all defects. The test may be made more sensitive by being carried out in a chamber or room from which both natural and artificial illumination are excluded.

Lines indicate cracks, seams or fissures running longitudinally in the direction of rolling or other working process, and only partly closed by these operations. They are caused by gasholes directly below the metal surface that have been crushed and oxidised. Their technical name is 'rokes'. Porous zones are indicated by a range of spots. If the indications are marked, any additional study can be carried out with a lens.

The indications are limited to surface defects. Those slightly below the metal surface are not revealed, and the test is primarily valuable for flaws on the surface, always liable to act as stress-raisers. Any metal, whether magnetic or not, can be tested, as well as such materials as tungsten carbide, and ceramic tools.

The penetrant used is chosen according to the object, and the main considerations governing its use are absence of fire risk and poisonous emanations. A large number of efficient penetrants are available and have been passed as safe by suitable bodies.

There is a somewhat different method in which the penetrant includes a powerful dye. When the surplus fluid has been removed, a 'developer', generally looking like a white pigment, is put on. The dye percolates into the developer and reveals important defects of the same type as shown by the fluorescent method. There is yet another test in which the fluorescent substance is combined with a normal magnetic ink capable of detecting defects. This test is primarily for examining magnetic metal components under the normal magnetic flaw detector, the components being examined under ultraviolet light.

The developer for the first of the two above tests is in the form of a thin powdery film serving to absorb the penetrant like a piece of blotting paper. The powder is usually caused to adhere to the metal surface by heating it until it forms the necessary film. The more usual penetrants include those with a base of oil emulsified by water and capable of being removed later by spraying under pressure with water. Alternatively the penetrant may itself dissolve in water and have a base of solvent or of oil using no emulsifier. The period of time during which the penetrant is allowed to act may be from $\frac{1}{2}$ to 30min, according to the work and the type of defects for which search is made, a longer period being required for close fine cracks than for wide ragged ones. Welds are best examined by brushing the penetrant on at the required zone. Big castings and forgings are usually sprayed or painted with the penetrative medium. Heating should not be attempted as its only effect is to drive off the penetrant as the temperature increases, no additional sensitivity for the test being achieved.

Either water or a solvent can be used to remove the penetrant, and all of it should be eliminated, none being allowed to come into contact with the piece thereafter. Developer may be of colloidal type in suspension, in which case it is applied to a wet metal before drying, but if it is dry and powdery, the piece, having been washed in water, is then dried at once either by heat or by a cloth, though heat is better. The developer is usually allowed to act for several seconds or minutes, according to the kind of defects, but for some big, coarse cracks a developer is not required.

The work should be studied for preference by black light, and the smaller the flaws, the less light of white or visible type should be allowed. Even fully dark rooms may be used on occasion. If there is any chance that the metal surface may be soiled or contaminated in some way, it must be well cleaned before the penetrant is applied, as otherwise percolation may be restricted.

The test is applicable to aluminium, magnesium, brass, bronze, ferrous metals, tools, welded pressure vessels, parts of automobile engines and seams in tungsten leads for vacuum tubes.

The operator must be carefully trained to use the test and also to interpret its results.

ELECTRICAL TESTING METHODS

Electric current flowing in a conductor always has a magnetic field associated with it, and in converse fashion a conductor moving in a magnetic field or a static conductor situated in a variable magnetic field has an electric current induced in it. These effects have been made use of in the testing of metals. Sensitive measuring instruments, such as the cathode ray tube, have helped in the development of electrical tests capable of giving greater efficiency at lower cost where ferromagnetic metals require to be inspected. One of the principal methods of this kind is 'magnetic analysis'.

MAGNETIC ANALYSIS

If an AC current flows through a coil wound on a tube of insulating material, an alternating field is produced of great strength inside the tube. When a metallic object is introduced into the tube, eddy currents are induced in it by field variation. Moreover, when the object is magnetic, these effects are highly complicated because the magnetic properties of, for example, iron, change in accordance with the field strength in which the object is placed.

Usually the metal in the coil alters both the field and the current passing through the coil in measurable ways. Ingenious instruments associate specific effects with specific metals or conditions of the metal, such as hardness, heat treatment, purity or work hardening. The main need is that there should be consistent correlations between the conditions and the effects. First, the condition or specific type has to influence the magnetic properties of the metal. Next, the variation in magnetic properties must distort the alternating magnetic field employed. This distortion of the field or of voltage induced by it must be measurable and enable the examiner to detect the condition producing the alteration in magnetic properties.

The test is applied to bars, tubes, small forged parts, castings, machined pieces and whatever has a uniform cross-section.

MAGNETIC ANALYSIS 131

Brittleness of the metal, its form, depth of case and ability to be machined, all have a relation to the stresses of mechanical type existing in the metal, and their magnetic effect. It is essentially a surface test and is not suitable for the detection of hidden internal defects, since the magnetic fields do not penetrate deeply. The minimum depth for detection ranges from 0·381 to 0·305mm (0·015 to 0·012in) according to the condition of the ferromagnetic metal. Mains frequency alternating current is used.

Of recent years this method of inspection has benefitted by the introduction of highly sensitive instruments and a greater power of resolution, with the result that measurements of wave form distortion are being extensively employed in commercial production testing. This method depends on the distortion of the wave form when ferromagnetic metal is introduced into the field, different metals giving different wave forms. Specific wave forms are not normally correlated to a particular property or condition, but as against this, many variations of metallurgical, mechanical, chemical or physical type influence only narrow sections of the fundamental wave cycle, so they can be measured by study of this cycle, even if the cycle is comparatively short.

Electronic point marking enables the instantaneous amplitude at points chosen at random in any induced voltage wave to be measured by circuits having gated vacuum tubes. The measurements can be at any part of the voltage wave, and this form of the test is employed for distinguishing qualities and solving problems of quality regulation in bars, tubes, and so on. A cathode ray oscilloscope helps to show up defects.

The rate at which the analysis is carried out is governed by the diameter and form of the metal, the ease or otherwise of its manipulation, the standard required and the skill of the operator. An average for reference is 20·32 tonnes (20 tons) of 2·54cm (1in) diameter bars in 8 hours. The apparatus may consist of a test coil assembly, a combination indicator and control cabinet and devices for working and feeding. The size of metal that can be dealt with is from 0·63 to 8·255cm ($\frac{1}{4}$ to $3\frac{1}{4}$in). The results are presented by individual meters or high speed indicators provided with neon lighting.

CORE LOSS COMPARISON

When frequencies above those of the mains are employed, the eddy currents and magnetic effects have even less penetrative power, until at extremely high frequencies virtually only a shallow case is affected. The mains supply has a frequency in Britain of 50 cycles/s. By introducing a comparator into a special double-wound coil so that it constitutes the 'core', the magnetic coupling between the two coils is affected according to the properties of the metal. Additional examination is given by the current frequency passing through the coil.

This test is specially suitable for the study of surface defects in case-hardened, decarburised, nitrided, and superficially stressed metals, as well as those season-cracked as a result of corrosion and internal stress combined, usually found in heavily cold-worked metals, particularly brass, and often caused by the presence of ammonia or its compounds. It is extensively used for weeding out the bad from the good metallic parts where hundreds of thousands are involved and the batch would otherwise have to be rejected *in toto*.

The principle of the test is that the metallic object, introduced into the coil field forming part of an oscillator circuit, causes a decline in the energy generated by the oscillator. These losses are of two kinds, magnetic and eddy currents, and in combination they constitute the entire core loss of the coil. The oscillator is extremely sensitive, but stable enough for effective results. The frequency range is from 1,000 to 200,000 cycles/s. The amount of change that can be detected is governed by the sensitivity of the oscillator.

The greatest advantage of this test is that it can be given directly following upon a critical stage in production, so that a check can be made before an operation, such as heat treatment, has gone too far to be corrected. The test is, however, of value also in non-destructive inspection at high speed of large quantities of bolts, bearings, small castings and stampings, while it is also useful as a means of sorting out sound from unsound parts in large batches previously jumbled together. Another use for the test is to reveal the start of plastic deformation.

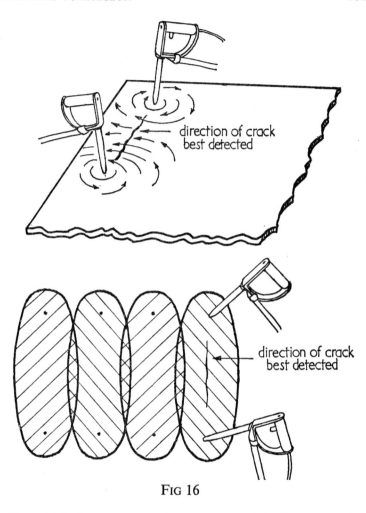

FIG 16

Since high-frequency core losses produced in the manner indicated earlier are modified by stress, so that these losses enable stresses to be compared when ferromagnetic metals are cold-worked or quenched, the extremely high residual stresses liable to produce cracks or early breakdown can be observed.

Some defects in metals, such as cracks, can be detected by a sharp modification of the eddy current losses reflected by the reading. This necessitates a special design of the equipment. The metal is rotated in a constant magnetic field and a small fixed pickup coil directly at the side of the rotating part develops a rush of current whenever a crack or defect goes past. This indication is amplified by tuned auxiliary circuits, showing that a crack exists (see Fig 16).

Radio-frequency crack detectors also use the differences in core losses to indicate the presence of cracks and are used when a considerable number of pieces need to be inspected.

CURRENT CONDUCTION

When a current passes through a bar or tube, the voltage falls from one end to the other by reason of the metal's resistance to the current passage. Defects or non-metallic inclusions considerably increase this resistance and consequently the fall in voltage, which is easily measurable. The object tested is, perhaps, in plate form, whereupon two 'prods' can be located on the surface and the voltage fall between them can be measured. Good results are obtained when tanks have corroded internally, and the interior cannot be reached without extreme difficulty, if at all.

On the other hand, eddy currents can be induced in the work by a small coil scanned over the surface. For tubes the coil (often known as a 'mouse') can be shoved from end to end of the tube to detect corrosion. The test is in reality a type of magnetic analysis, but is of great service in the examination of condenser tubes in for example ships and marine power stations.

Chapter 8

CORROSION TESTS
The true test of a metal's resistance to corrosion is its life in service. All other criteria are essentially approximations to the truth, never the truth itself. Since it is virtually impossible to repeat in a laboratory or research station the precise conditions of corrosion under which a metal will be called upon to function, any test carried out in these institutions can be but a pointer. The best approximation to actual service is a field test in a comparable set of conditions and, if possible, in an identical or largely similar installation. Such tests are often carried out in a small pilot plant.

Ingenious corrosion tests have been devised, particularly for ferrous metals. They include the salt spray, total immersion, alternate immersions, accelerated electrolytic, atmospheric and gas tests, etc. Specimens of varying design and form have been adopted, but interpretation of the results of these tests must be based on the field tests or tests under actual service conditions. It is difficult to establish quantitative data, and the results are invariably affected by the temperature and degree of aeration of the corrosive medium, the condition of the metal surface in advance of the test, the duration of the test, the methods used and the attention paid to cleaning and drying the tested piece before weighing.

The Salt Spray Test The general principle of this is the careful weighing of a sample of the metal, and its introduction into a carefully sealed chamber or container. A fine spray of a saline solution, consisting of about 20% of sea or common salt + water, is continuously injected into the chamber for a considerable period in the form of extremely fine drops constituting a fog or mist. The normal test lasts about 100h. The container or chamber is made of a corrosion-resistant metal or material, and the air whose pressure produces the fine spray is carefully filtered to ensure that it is free from all particles of foreign matter that might affect the test results. Care is also taken to prevent any change in the spray composition such as might result from a change in the moisture content of the incoming air.

The mist must not condense so that droplets fall upon the testpiece, nor should the droplets formed by condensation impede the spray. The test is intended to reproduce as far as possible the effects of a marine atmosphere on the ferrous metal, and the direct descent of droplets or a flow of saline water would invalidate the test. Recirculation of the spray is likely to affect the rate of corrosion owing to the products of corrosion being circulated; therefore this is not recommended.

The test data are expressed as penetration/hour or other period of time, and the measurements taken with great regularity.

The spray solutions may not always be saline. Some may be of city or mine water or of acid type, designed to simulate the conditions of service. Sometimes they may be of distilled water plus an added substance to prevent any possibility of impurities affecting the tests. The results of the immersion test are expressed as weight lost, physical properties affected or the effect on the surface in the form of pits and their depth.

It is always a good plan to test not only the metal concerned, but also welded testpieces of the kind likely to be embodied in the particular construction. In these instances the weld deposit is tested.

Immersion Tests In one of these a testpiece is subjected to a damp atmosphere corresponding as far as possible to that of service and so controlled that a predetermined standard composition is maintained. The testpiece is allowed to remain for a specific length of time in artificial light, after which it is washed, dried and exposed again, this cycle of operations being constantly repeated until a measurable amount of corrosion has been produced.

The full immersion test is more stringent and involves hanging the testpiece by a non-corroding hook of some such material as glass in a corrosive bath representing the agency liable to attack the metal in service. The conditions of service can be controlled as regards composition, temperature, type and pattern of motion in the liquid medium. In similar tests of this kind short testpieces may be suspended from a planetary gear which rotates and drives them through the corrosive agency at a predetermined rate.

Field Tests These subject the testpiece to service conditions and locations representative of the eventual working conditions for the finished part. The place and conditions of the test must not change, as this renders the test nugatory. There are two different service tests of field type. In one, a testpiece is used, in the other a complete component. The second of these is more expensive, but better. For example, either a testpiece of small dimensions or a complete baffle plate may be introduced into the chimney stack of a furnace. Similar field tests are essential when the behaviour of a metal under soil corrosion or mine corrosion conditions has to be evaluated. There are a great number of varying conditions for different localities, and the electrochemical action is by no means easy to reproduce except under a field test employing identical conditions.

A field test of considerable utility is that in which a number of pieces or sheets are exposed to an industrial corrosive, a country and a marine atmosphere, so that the different corrosion rates of three quite separate atmospheres can be determined for the particular material. Industrial atmospheres are far more corrosive than the other two types, since they contain soot, sulphur compounds, dust, gases, chlorine compounds, grit, and so on. These are transported by raindrops, and their deposition on the surfaces of the sheet tested soon starts a degree of corrosion.

Metal corrosion is perceptible by discoloration of the testpiece, weight loss, pits or depressions, and hidden corrosion below the surface, shown by a decline in strength, and modification of other mechanical properties.

Discoloration is not necessarily important except as regards the ornamental appearance of the metal, and indeed, the fine green patina of copper sheets on buildings is often an attraction. Nevertheless, it is advisable to test what is happening below the surface and possibly causing deleterious effects. A specimen is polished after cleaning and for this cleaning benzine may be used. The likely corrosive agent is dribbled on to the testpiece for a brief period, and after a time the piece is examined to determine the amount of discoloration.

Loss of weight is serious and may be ascertained by regularly weighing the testpiece and showing the loss in mm of penetration/year. Loss/year declines with time in most cases. In many instances the testpiece is given a short corrosion time in the particular agent in advance of any weighing, the object being to make the surface uniform throughout.

Pits are measured by a depth gauge and expressed as a corrosion penetration in mm/year. These pits are ready initiators of corrosion cracks, so that this value together with the weight loss give a fair indication of the likely resistance of the metal to corrosion.

The hidden form of corrosive attack is termed 'intergranular corrosion' and takes place at the boundaries of the grains. No direct measurement of this is possible, so that the amount of susceptibility can be decided only by a comparison of initial tensile strength, elongation percentage, electrical resistivity and other properties, before the test and at intervals during it. If any of the more important properties show a marked decline without any visible change in the surface condition, it is fairly certain that intergranular corrosion is responsible. Regular examination of the testpiece under the microscope is therefore advantageous.

Some metals age, that is, their mechanical strength changes without apparent cause over a period of time. This must be taken into account when evaluating the tests.

Accelerated Corrosion Tests These are intended to indicate in a short time what a service test would require months and even years to tell. They are never so satisfactory as service tests, which cannot be imitated fully in the research laboratory.

In one such test a testpiece of the right thickness is taken and welded on to another by the electric arc process, the second testpiece having a higher carbon content. The two are then immersed in a bath of nitric acid (10%), hydrofluoric acid (2%) for 24h at 60° C (140° F), the solution being changed at half-hourly intervals. A more rapid test is immersion in a bath substituting hydrofluoric for nitric acid, using a 50% solution at 100° C (212° F). This solution, too, is changed at half-hourly intervals.

FLUIDITY TESTS

These are tests to establish how far a metal will flow into and fill a mould without solidifying before it has reproduced the desired form in all its detail. It is a simple means of ascertaining the viscosity of molten metal, but it differs from low temperature viscosity tests of fluids such as oils, etc.

Metal fluidity is governed by viscosity, surface tension, surface films, gas content, non-metallic inclusions, and the way in which the metal freezes and crystallises. The testing methods cover such variables as the shape of the testpiece, the size and form of the casting head, the type of moulding sand used and its characteristics, pouring rate and the temperature of casting.

There seems to be no standard form of testpiece, and the test itself usually involves the production of a channel of small cross-sectional area adequate to ensure cooling to solidification and cessation of metal flow. The basis of the test is the distance between starting of flow and finishing, and it necessitates a specific relation between the cross-section of the channel, the force required to propel the stream along the channel, and the heat-resistance of the moulding sand. Most channels in this test are either spiral or straight, and in form semi-circular, circular or trapezoidal, the cross-section being 53·8 to 92·2mm² ($\frac{1}{12}$ to $\frac{1}{7}$in²). The metal must flow at a uniform rate through the channel and at a constant pressure, and the head through which it is poured must be carefully controlled. The mould should have adequate permeability so that all the gases given off as the result of hot metal meeting moist sand are voided, and the figure suggested is 75. Adequate venting of the mould is essential and the consistency of the sand should prevent erosion by the hot flowing metal. These requirements are met by careful choice of moulding sand grain size, washes and surface coatings.

However carefully these tests are carried out, some testing should be done in the foundry itself, as there is no proven correlation between the tests outlined here and the way the metal will behave in production. The tests depend on many variables and are not standardised, so that any departure from strict observance of the requirements specified may nullify the values obtained.

Fluidity in any event should not be regarded as a synonym for castability. Castability can be improved by careful attention to fluidity.

In steel melting, the slag plays an important part, as in the acid electric arc melting process for example. Here the fluidity of the slag is taken as an indication of the oxidation throughout the process. Measurement of viscosity is used as a quantitative test in many instances, and an active oxidising slag is one with an 11 in value for a melt with about 40% silica and iron oxide about 41%. Fluidity of the slag declines as the boil continues and slag acidity increases. Towards the end of the operation fluidity decreases, and readings of 2 in viscosity signify a silica content of about 56%. A special type of viscometer is used for this test.

TESTS OF MACHINABILITY

One of the most important properties a metal possesses is the comparative ease with which it can be turned, drilled, milled, threaded, broached, sawn, reamed, etc. The user or machinist requires to know something of the speeds and feeds at which metal can be removed, the best method of getting the most economical service life for the tools used when put to work, the power the machine tool will have to exert to drive the tools at the most effective cutting rates, the kind of surface finish the machine will give on this particular metal, and sometimes the extent to which accuracy to size can be achieved and the components produced to identical dimensions.

Numerous kinds of tests have been devised to measure machinability. For such tests to have validity, the material cut must be as far as possible identical in form and properties for each and every test. Similarly the form, cutting angles and composition of the cutting tool, as well as its type, for example high speed steel, cutting alloy (Stellite, etc), tungsten or other carbide, or ceramic, must remain the same. The type of coolant or cutting fluid used may constitute a variable factor if desired.

Under these conditions, the test should show the service life of the tool in the specific cutting operation, the cutting speeds and feeds required to cut at the most economical rate, the power required for the cutting operation, the depth of cut possible for the given speeds, feeds, etc.

The data obtained can be expressed in various ways, such as the rate at which the tool will drill or cut or otherwise machine at an unvarying feed pressure and in a specific period of time. Another measure is the traverse or sideways movement in the lathe a cutting tool makes during feeding under constant pressure; and again the test may be expressed as the time in min or number of cuts necessary to cut off a specified cross-section under a specific power feeding load instead of a positive feed rate.

Another test may be the finish of the surface after completion of machining or the amount of frictional heat generated at the tool nose and its effect on the cutting power of the tool. In most machine shops, however, the usual criterion of a properly conducted test is the expense involved in machining away $16 \cdot 39 cm^3$ ($1 in^3$) of the metal, given the specific conditions required. The advantage of this form of expression is that it covers not only tool and labour, but also power cost.

Some users employ machinability ratings or indexes to enable them to choose the right conditions for machining a type of metal not previously encountered, but these do not give a true basis for machinability as they omit important considerations of expense, while the rating of a material established by one method may be quite different from that obtained from another. Usually the only true way of comparing the machinabilities of various metals is by establishing the best condition for economical cutting, and then comparing the performance of each type or tool or metal under these standard conditions.

TESTS FOR RESISTANCE TO OXIDATION

The ability of a metal to withstand oxidation at elevated temperatures is mainly established by methods depending on the purpose for which the metal is required and its physical form. These methods include ascertaining the increase in weight of a testpiece subjected to the oxidising conditions (a) for a particular length of time; (b) after elimination of the oxide scale formed; (c) by measuring the quantity of gas used in producing the oxide; (d) by establishing the time necessary to destroy a specific size of metal; and (e) by metallographic study of the oxidised metal.

These methods all show varying properties of oxidation and the results are entirely governed by the test given. Most popular are those tests that decide how far the metal can develop an oxide scale low in electrical conductivity and how far the refractoriness or heat-resistance of the scale produced will enable the metal to withstand oxidation.

All the tests enumerated are useful, but how far a particular metal will be stable in an oxidising atmosphere cannot be established with certainty. The tests can only be regarded as comparative for specific heating and cooling ranges and cycles, furnace atmospheres and additional circumstances. A guide is the temperature below which oxidation loss is less than 0·0002 to 0·0004g/cm^2/h (0·03 to 0·0g/in^2/day.

TESTS FOR RESIDUAL STRESS

A residual stress is one locked up in an elastic solid body when not subjected to load or other external stresses. It may be set up by cold working, such as stamping; a modification of the specific volume as a result of expansion under heat or contraction in cooling; by electro- or magneto-striction; or by the union of metals by power, as in welding.

It is not easy to discover the size and direction of these stresses. Various suggestions have been made, in one of which the metal is drilled, the alteration in strain produced by the relief of stress at the free surface being determined by strain gauges. This is not an accurate test, however. An alternative is to trepan a plug with a hollow tool and ascertain the alteration in strain on three or more axes on both plug surfaces. This is also inaccurate if the residual stresses are not consistent over the whole plug.

A third method is to use X-ray diffraction employing back reflection, but this measures surface stresses only. No trustworthy non-destructive test other than this has, however, been invented that will indicate residual stress over the entire piece. Hence, it is best to employ established stress relief treatments, such as thermal or mechanical.

TESTS FOR SEASON CRACKING

This combination of corrosion and internal stress arises when metals are heavily cold-worked. There are various ways of discovering whether a metal is susceptible to the defect. One of the easiest is the immersion of the metal, which has first been thoroughly cleaned, in a mercurous nitrate solution for a certain length of time, after which cracks, if visible, show that the metal will fail in service. The solution contains an acidifier of 1% HNO_3, and either 1 or 10% $HgNO_3$. The immersion time ranges from 0·25 to 1h. If the metal withstands a 0·25h immersion it is not likely to show season cracks in normal circumstances, but although trustworthy up to a point, the test is not absolute.

An alternative test is for brass, and comprises exposure of the metal to ammonia gas, water vapour and air combined for from 4 to 24h. This is not a widely accepted test, is expensive in equipment and cannot be correlated with cracks in service.

In a different type of test, mechanical rather than chemical, the stresses in intricate forms are determined, the dimensional changes occurring when small successive increments of metal are removed being measured by various means. The tests become less and less accurate as the object becomes more intricate, and not only take up much time, but are also costly. They also fail to show up concentrated or localised peaks of stress, though they do enable values to be determined.

Yet another method is that of X-ray diffraction of an intricate kind. The advantage of this is that it is non-destructive, but measures only the stresses in the region studied.

STRESS CORROSION TESTS

Accelerated corrosion tests are not to be trusted for stress corrosion unless the test can be correlated with the actual service behaviour of the metal in question. There is no difficulty in producing cracks in metals of almost any kind if they are subjected to severe stresses and immersed in carefully chosen solutions, but this does not means that they would necessarily show the same susceptibility to cracking under service conditions.

Stress corrosion is a deterioration in the mechanical properties of a metal, through the combined action of static stress and a corrosive atmosphere or agency, greater than would follow upon the individual effect of these factors. The tests are usually on testpieces previously stressed in a corrosive solution or medium to establish their susceptibility to attack. They are usually performed under either constant stress or constant strain.

Corrosive attack in particular regions results from the anodic relation of these regions to the rest of the construction, and the cracks when they occur are caused by electro-chemical action between anodic and cathodic zones.

Aluminium is not sensitive to stress corrosion cracking, but these high strength aluminium alloys normally heat-treated are susceptible in certain conditions of heat-treatment, but as these are known the cracking can usually be prevented in use.

TESTS OF WEAR

These are of increasing importance because the conditions of modern life impose ever-increasing rates of wear on metals. There are many definitions of wear, one of which is 'the slow abrasion of surfaces in friction contact'. A more recent one is 'the process by which metal is removed from surfaces moving in contact with one another, as in abrasion'. An American definition is 'the gradual deterioration by mechanical means of surfaces in contact, as by tearing off particles through friction, but it may be modified by corrosion or other chemical attack occurring simultaneously'.

The measurement of the wear resistance of a metal becomes, therefore, increasingly important, but it must be realised that no all-embracing test is acceptable, since wear is of many different kinds and the test for one kind may result in injury to the properties of a different metal. The first requirement in wear testing is that the data obtained must be correlated with the results in actual service.

The considerable number of mechanisms devised for the purpose prevent detailed description of each and every one, so that here we must confine ourselves to the main types, which can be divided into two groups; tests in which one metal rubs against another; and those in which the metal rubs against a nonmetallic substance or against a material expressly designed to abrade it.

The various machines all endeavour to reproduce an approximation to the wear that will be encountered in service. Some employ an abrasive wheel or a revolving abrasive disc, but whatever the method adopted, the constant factors must be the loading conditions, the speed and the lubrication employed. The variables are many, such as pressure, temperature, speed, surface finish, lubrication, type of abrasive, etc. Hence the necessity of choosing at least some constants from this great variety. The results of the tests must, in any event, be carefully evaluated, as the omission of but one factor may invalidate them.

The primary objective is to devise a test that will simulate actual conditions of service as closely as possible, but this presents many problems. No attempt should be made to accelerate the test by short cuts, which usually take the form of a magnification of the conditions of actual use. Such attempts merely upset the balance in other directions. Higher pressure, for example, may result in too great a surface hardness or distortion of the metals involved. Too high a working temperature may follow on higher friction.

A good plan is to plot the data from continuously-following tests against time, total travel, or some other factor. The curve obtained will be governed by the surface finish of the metal. It is also possible to compare the ratio of weight loss to that of a chosen 'standard' metal working in conditions resembling those of the actual wearing operation. This is a useful method where the two testpieces can be tested at one and the same time and it is not practicable to standardise the conditions of the test. It means that both testpieces undergo the same changes in conditions. The method is, however, less direct than the curve method earlier described.

It is essential to maintain close control of the test conditions, but whatever test is used, it must place the various metals in order of precedence, and this order must coincide with that shown by the same metals in working conditions.

Oxide films on metallic surface are also liable to affect the test results because with some metals they prevent wear to some extent.

Apart from a few exceptions, the most important factor which governs wear resistance is hardness, at all events for steels, but the ordinary Brinell and Rockwell hardness tests are not of much use here, since the data they give are inadequate and may cause the investigator to go wrong. Much more trustworthy is microhardness testing which discloses facts not recorded by alternative methods.

The best plan of all is still, however, to choose the metal to be tested, form it into the right shape and dimensions so that it becomes an integral part of the working machine in which it is to be used, and study its wear resistance under actual service conditions. Should it give good performance it may then be made into a production model and tested for a longer period in more variable circumstances, careful records being kept of service life and performance.

Wear in relation to surface finish has been measured by the establishment of a relation between the finish and the load necessary to score the metal. A scoring machine for the laboratory reciprocates a scoring tool at a rate of 609·6m/min (2,000ft/min) when properly lubricated. The smoother the surface finish, the higher loads the metal will withstand.

TESTING PERMANENT MAGNETS

Magnets that retain their magnetism after removal from an external magnetic field are usually tested for quality by measurement of certain fundamental properties, such as remanence, coercive force and the maximum amount of magnetic energy a permanent magnet of a specific metal will withstand in an air gap. These are respectively abbreviated to B_r (sometimes B_{rem}), H_c and BH_{max}.

The instrument employed for testing is a permeameter, which measures one magnetic permeability of a cast or rolled properly heat treated bar material. The magnet in its finished form is usually given a flux test for which an exploring or 'search' coil is used, being a small coil for measuring the flux in a magnetic field. This is accompanied by a flux meter of galvanometer type, and the two instruments in conjunction provide the maximum accuracy. For commercial purposes a less-accurate test is required which is considerably quicker than the previous test. The practice is to use a DC flux meter, of which the magnet provides the field causing the moving coil of the meter to deflect. On the other hand, it is also possible to adapt an AC meter of the iron vane pattern, while voltage measurements from air core coils rotating in the air gaps or fields of powerful magnets have been used.

The tests outlined above have all been used for both magnets and metals.

TESTING ENGINE VALVES

The valves used in automobile engines must not lose strength at the high temperatures of service or they will fail. This means that as far as possible their performance must be predetermined by suitable tests.

One test that has proved useful in this respect is the creep rupture test, which involves heating a series of valve metals and/or alloys to 700° C (1,300° F) for a period ranging from 100 to 1,000h according to the material.

A tensile test of brief duration is sometimes employed, especially if creep data are not at hand, while hot hardness tests are also used, one of them being a common indentation test in which two cylinders of the metal are forced into contact at suitable temperatures by a predetermined load. The surfaces where contact takes place are necessarily flattened, and the flat zones are measured. This test is primarily for exhaust valve metals. The Brinell test is used to determine hot hardness only when no difficulty is encountered in using it.

The behaviour of the metal as regards fatigue at high temperature is also important owing to the severe alternating stresses of service. High temperature compression-tension tests are employed.

LEAD CORROSION VALVE TESTS

The use of lead in the fuel of the automobile is a constant factor in the corrosion of such parts of the engine as the exhaust valves. The seating faces, the underhead and top part of the valve stem, and the upper portion of the head, are the points most susceptible to either corrosion or undesirable deposits. This has made it essential to devise some form of test to determine beforehand how an alloy for exhaust valves will behave in service. The usual test adopted is known as the lead oxide crucible test.

In this a testpiece of about 11·03mm (0·444in) in diameter by 11·03mm (0·444in) long is inserted into a pot or crucible of magnesia together with about 40g of lead oxide. The pot and contents are then raised to a temperature of about 910° C (1,475° F), allowed to cool to room temperature, carefully cleaned, then weighed. There will be a loss of weight, which when measured indicates the comparative corrosion resistance of the alloy. The test is solely for lead oxide, but has been found suitable for use with automobile engines.

VALVE ELONGATION TESTS

It is sometimes found that the length of a valve used in automobile engines is extended permanently, probably because a slow cupping of the valve head occurs. This necessitates an alloy for the valve capable of withstanding a high temperature without marked loss of strength. The unchecked extension causes in the end imperfect valve seating, exhaust gases escape and play on the valve, which eventually fails.

A special elongation test is applied to the valve materials to determine their suitability for this form of service. The conditions of service are imitated for 100h in light duty engines. Valves made of austenitic 18–8 chromium nickel steel are compared with valves of steel containing a hard, brittle, non-magnetic compound of intricate composition known as 'sigma phase', which undergoes severe embrittlement after specially long exposures within a particular range of temperature.

If the result is a negative elongation, this indicates that as the engine rises in temperature, clearance between valve tip and tappet is probable. Sodium cooling is often successful in minimising this clearance.

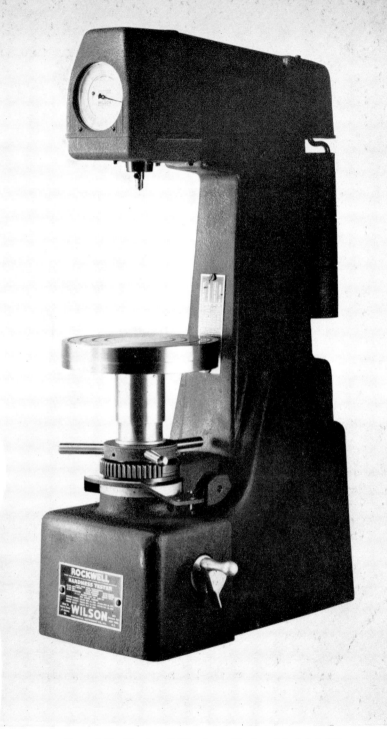

Page 149 Rockwell Hardness Tester, Model 4–JR

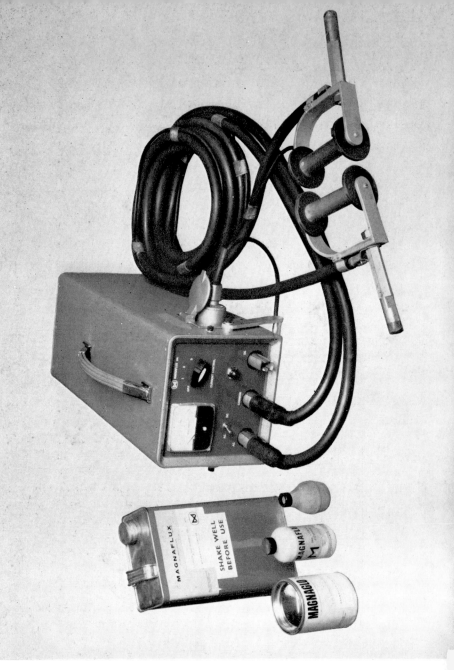

Page 150 Magnaflux Type KH 12 ac/dc portable magnetic testing unit

GAMMA RADIOGRAPHY TESTS UNDER WATER

Ocean-going vessels are frequently welded below the water lines, and as opportunity offers it is a sound plan to have the welds tested. This involves the employment of under-water divers who have to work in conditions of cold and difficult vision through muddy or otherwise dark waters. The method adopted is to prepare beforehand working drawings of the hull on which the chosen stress zones are coded. A technician is located in the right place within the vessel and carries a radiation source embodying a radioactive isotope.

The diver determines the position of every butt weld by touch and marking with a magnet, a magnetised film holder being set against the marked positions and an exposure being made for each weld. Gamma rays are used, and as many as forty examinations of this kind may be required for the one hull.

TESTS FOR MOLYBDENUM'S EFFECT ON WHITE IRON

When an iron is required for resistance to abrasion, the most effective resistance to wear is obtained when the iron is alloyed with 1 to 4% molybdenum and has a pearlitic matrix, especially if 5% chromium is included in the composition, and in particular if the alloys are of 12 to 18% chromium type.

To test the white irons so alloyed a range of testpieces not of standard type but measuring 30·48mm (1·2in) diameter, are cast in a sand mould. On removal from the sand at a temperature of about 980° C (1,800° F), the testpieces are cleaned with a wire brush, then introduced into a furnace at this temperature. Here they are left for 10min and end-quenched on removal, that is, the quenching is at one end only, as in the Jominy End-Quench Test, the hardness being determined at previously chosen intervals longitudinally from the quenched end.

A curve is then drawn giving the results of these determinations, and a point on this represents the distance from the quenched end marking the appearance of pearlite.

K

ROLL HARDNESS TESTS

A roll is required to be hard so that the surface will not disintegrate under heavy pressure, and the roll can work long enough to achieve maximum economy. This involves measuring the hardness, which is often carried out with the scleroscope, a portable hardness-measuring instrument which quickly establishes a hardness value without indenting the roll surface to any significant extent. On the other hand, the Rockwell C test is adaptable to small diameter rolls which can be carried to it. It is possible to correlate scleroscope with diamond pyramid hardness values by suitable charts.

DIE BLOCK TESTS

Many steel manufacturers stock for their customers blocks of die steel suitable for a particular purpose. These have already been graded according to composition and given a heat treatment designed to provide adequate hardness. The blocks are usually tested for hardness by the Brinell machine employing a tungsten carbide ball indenter.

The reason for this is that the Brinell test is not responsive to the small constructional differences between blocks of considerable dimensions, and also because the Rockwell machine is not suitable for tests on such large pieces. The tungsten carbide indenter is also less likely to deform under the pressure of indentation with a block of considerable hardness, and maintains its dimensions more consistently than the ordinary steel ball.

The scleroscope can also be used since it can be taken to the block instead of the reverse.

TESTS FOR STEEL DAMAGE BY HYDROGEN

Steels used in petroleum refinery components are often subjected to the action of hydrogen at high pressure and temperature.

This may result in a destructive weakening of the metallic surface, resulting in cracks produced by intercrystalline corrosion. These cracks cannot be seen with the naked eye, and rapidly reduce the mechanical strength of the metal. Trustworthy tests are essential to determine if any regions of the component have been attacked in this way.

While some data can be obtained from bend or flattening tests, the most trustworthy is the ultrasonic test, which by its character does no injury to the component, but clearly indicates those areas that may have been weakened. These can then be isolated and given additional examination. How far the effect has penetrated is by no means easy to determine, but in such instances the ultrasonic test can be supplemented by placing the specimen in a bath of hydrochloric acid (50%) at about 80° C (180° F) for 20min. Examination then reveals any zones that have suffered damage.

These hydrogen-embrittled steels often crack as a result of this attack, so that a good rough test is their hardness, which is usually lower than normal, or indicates a microstructure filled with minute gasholes. Tensile tests or close study under the microscope may also help to determine the affected zones.

TESTS FOR STAINLESS STEEL QUALITY

Users on many occasions have serious problems of identification of stainless steels, either because the distinguishing marks have been obliterated, or because miscellaneous compositions have been lost in stock warehouses, etc. While it is always possible to determine the types of stainless material by chemical testing in the laboratory, this takes up a great deal of time which cannot always be allowed, while another deterrent is the expense.

It is necessary, therefore, in many instances to devise a rapid and simple test that will not, perhaps, give a detailed answer to the question of composition, but will readily determine whether a metal is indeed a stainless steel, and also whether it is of ferritic or austenitic type.

The most convenient and quickest test of this kind is the magnetic. By passing a magnet over the bar or piece concerned, it is easily seen, if it suffers no attraction, that it cannot be a ferritic stainless steel, which is magnetic. The test is not absolute, because some severely cold worked stainless steels may acquire a minor degree of magnetism, but it does prevent the user from mistaking the austentic 18–8 steels for the plain chromium-carbon or low alloy steels. It should be noted, however, that nickel silver is also non-magnetic.

Another rapid test is the nitric acid one, in which an application of concentrated nitric acid is applied to the doubtful metals. The true stainless steels should easily withstand attack by this, but the martensitic stainless steels containing 12 to 14% and 16 to 18% chromium without nickel may show a certain small attack, while non-ferrous alloys readily betray the effect of the acid. If the distinction is to be drawn between stainless austenitic steels and ordinary carbon steels, it is not necessary to employ concentratated nitric acid, as a dilute solution will strongly attack the carbon steel.

A form of heat-treatment constitutes a good method of distinguishing between those stainless steels which show a slight amount of magnetism after severe cold working and the true non-magnetic stainless steels. The specimen of metal should be heated to between 1,010 and 1,090° C (1,800 and 2,000° F) and plunged into water, whereupon the true stainless steels will be without response to the magnet and show a hardness of Rockwell B85 (maximum). Other steels of different composition can be separated from each other by heating them to between 950 and 1,010° C (1,750 and 1,850° F), when 14–18 and 25% chromium steels are softened to below Rockwell C24 and if in the form of wire are readily bent by hand. Other stainless steels of martensitic types show a Rockwell hardness of C36 to 43, but the wire made of these steels cannot be hand-bent without difficulty.

The high carbon stainless steels similarly tested show a hardness of Rockwell C50 to 60. Table 3 will aid the reader in deciding the type of steel.

Copper sulphate provides yet another rapid test, involving the immersion of the doubtful metal in a bath of 5 to 10% solution termed 'blue vitriol'. The piece is cleansed entirely of dirt, oil, grease or other coatings, and the surface is rubbed with a soft emery paper in advance of immersion or application of the sulphate solution. If this latter method is employed, the solution falls upon the piece from a bottle specially designed for this purpose. The solution will give carbon steel or iron a blue tint very quickly, the stainless steel remaining unchanged.

TABLE 3
ROCKWELL HARDNESS

Composition	Rockwell Hardness after End Quench Test
Austenitic	B85 maximum
14–18 Cr, 0·12 C	Below C24
25 Cr, 0·2 C	Below C24
12·5 Cr, 0·15 C	C36–43
12–14 Cr, 0·15 C	C36–43
12·5 Cr, 0·15 C, 1·25–2·5 Ni	C36–43
15–17 Cr, 0·2 C, 1·25–2·5 Ni	C36–43
12·0–14·0 Cr, 0·15 C	C50–55
16–18 Cr, 0·60–0·75 C	C55–60
16–18 Cr, 0·75–0·95 C	C55–60
16–18 Cr, 0·95–1·2 C	C60 or above

Sulphuric acid provides a useful test, but in this instance complete immersion of the metal is essential. The test is usually carried out on specimens cut from sheet or other form, and finely ground on the edges, most effectively done by honing. Cleaning follows, after which a solution of nitric acid, 20 to 30% by volume, gives the metal a protective film, the specimen being immersed in this for 30min at 60 to 65° C (140 to 150° F). The sulphuric acid solution is heated to 70° C (160° F) and is 10% by volume. Soon after immersion, if the specimen gives off bubbles strongly, it is of the 18–8 stainless steel type, and after a short period the metal becomes dark in colour.

This test is limited in value to differentiation between these types of steel and other stainless types containing basically 16–18 chromium, 10–14 nickel, and 18–20 chromium, 11–15 nickel, %, which give off only a few or no bubbles and do not alter in coloration unless immersion lasts longer than 15min.

Cold acid of sulphuric type can also be used as drops allowed to fall on metal surfaces previously ground, abraded with a file, scrubbed or given a coarse polish. The surfaces are all tested, and this time the difference does not lie in presence or absence of bubbles or in a contrast between discoloration and none, but in the rate of attack and difference of colour. The distinction is between the same classes of steel as for the previous test, but this time the 18–8 steels behave as for sulphuric acid (20% concentration), but the other steels have a much slower reaction. The 18–8 steels turn dark brown or black and eventually produce greenish crystals on their surfaces. The other steels change tint only gradually, and at first acquire a yellow-brown hue, which eventually either turns fully brown or produces dark crystals with a faint green colour if the attack lasts long enough. The 18–20 chromium steel reacts at a more gradual rate than the 16–18%.

Phosphoric acid is sometimes used to distinguish stainless steels of the 18–8 and other chromium nickel stainless steels from similar steels with a molybdenum content. The immersion is in a warm solution of concentrated phosphoric acid and 0·5% sodium fluoride, at 60 to 65° C (140 to 150° F). The steels not containing molybdenum do not suffer attack, whereas the non-molybdenum steels gradually give off bubbles of gas.

Another acid solution for distinguishing between low chromium stainless steels and those with higher chromium is hydrochloric acid (50% by volume) used to dissolve equal weights of small chips of the metal to be tested. The various batches are ranged in depth of colour (green) with the higher chromium steels giving the deeper green.

A similar solution of 40 to 50% by volume can also be used to distinguish between austenitic and martensitic stainless steels. The method is to insert a few chips or drillings into individual test tubes, one for each steel, and add the solution. The austenitic steels will produce a pale blue green or light amber tint, the martensitic react strongly and turn a much darker green. Some of the austenitic steels give off a garlic smell, whereas some of the martensitic produce a rotten egg smell. A few of the austenitic steels also give a light blue green tint, but without any offensive smell.

Machining tests on stainless steels are primarily to distinguish between the free-cutting and the conventional types. The specimen or piece is turned in a lathe and the character of the turnings produced compared. The free-cutting steels produce chips that can be easily broken up and are brittle, whereas the conventional steels produce much more tenacious and springy chips. It is also found that when machining the free-cutting steels, any form of dry machining such as grinding or sawing causes an offensive smell. If further assurance is required, this can be obtained by microscopical inspection of the testpiece, which should be carefully polished and longitudinal.

Stainless steels, after polishing, can be pressed for $\frac{1}{4}$min against a sensitised paper previously saturated with a solution of 5% sulphuric acid, which on removal and inspection shows a dark brown tint if the steel is of resulphurised type. However, if the steel is so low in manganese content that manganese sulphide is, not present but chromium sulphide instead, no reaction on the paper will occur.

It must be clearly understood that all the tests described under the general heading of Tests for Stainless Steel Quality are merely rough and ready means of distinguishing some kinds from others, and depend for their success on the care with which they are carried out and the operator's understanding and knowledge. It is not suggested that a proper and complete test by metallurgical analysis can be replaced by these methods.

TESTING NODULAR IRON (DUCTILE OR SPHEROIDAL IRON)

This is largely carried out by examination of the microstructure of the iron, as exemplified by the standard drawn up by the American Foundrymen's Society. This uses a sand-cast testpiece typical of the melt. A series of known standards is then used for comparison. The specimens for microscopic examination are both unetched and etched. The unetched reveal the form in which the graphite is present and the size of the spheroids; the etched reveal microstructures. For heat-treated or stress-relieved iron, it is usual to prepare an additional range of annealed ductile irons having different established retained pearlite contents for comparison.

The first step is to test cast specimens by the normal tensile and other testing methods of destructive type. The specimens should be of different ranges of weight. The microstructures of these are then compared with those of the specimens taken from the melt, which must finally be compared with the mechanical testpiece values of both types of specimens. The time spent in between treating the iron with magnesium and teeming into the casting mould is then ascertained and gives a measure of the decline in the influence of the added magnesium. This usually begins after 22min, but may start as early as 12min. The actual time is established by teeming ladle samples at various regular intervals and comparing them with samples taken from castings.

TESTS OF BOLTS

Perhaps the most popular method of testing bolts is by a flow rate method devised by E. L. Robinson, who established the rates of creep by means of a range of declining loads chosen to regulate total elastic + plastic strain. General practice for creep testing has already been covered on page 55. Bolts are tested in a similar manner, but the limiting strain following on the initial load is regulated by the elastic recovery on reduction of the load. For a wide range of metals the approximate relation in creep test flow rate values is $r = r_o \left(\dfrac{S}{S_o}\right)^n$, r being the rate of creep at the stress S and r_o the rate of creep at the stress S_o. If this relation is actual, the metal will behave roughly in accordance with the following equation:

$$S = \left[\frac{bS_o^n}{(n-1)r_o Et}\right]^{1/(n-1)}$$

S being the stress after the time t, r_o the nominal rate of creep at the stress S_o, n the exponent of the equation for creep rate, E the elastic modulus at the appropriate temperature, and b the elastic follow-up of the system.

Since n and S values are governed by initial stress and elastic follow-up, the equation does not show how a metal will behave under a starting stress or elastic follow-up considerably different from those used to establish n and S_o. Moreover, the equation in question is founded on virtually constant elastic modulus.

The refinement of this test by later investigations involves complex equations and statistical analysis for which space is lacking, but what is known is that metals for bolts notch-sensitive in fracture have to be used in circumstances so strictly regulated that the mean stress is always considerably less than the notch rupture strength.

For high temperature work, bolts have the initial stress carefully controlled, the stress being of a specific value, and the range particularised as from 0·001 to 0·0025 with minute tolerances. To measure the strain an external micrometer is the best instrument.

TESTING ELECTRODES FOR MAXIMUM WELDING SPEED

Electrodes used in welding are often required to function at the greatest possible welding speed that will not sacrifice quality. For electrodes used in horizontal fillet welding lap joints on plates vertically positioned, manual testing proves inaccurate owing to the human element. In consequence a machine was devised in which an assembly was placed on a moving carriage which transported it at a fixed speed past a test electrode mechanically pressed into the groove at specified pressure. The carriage had a speed range of 177·8 to 5,333mm/min (7 to 210in/min), and the testpiece was secured to its side by an extension. The feed mechanism for the electrode was an inclined 15·88mm ($\frac{5}{8}$in) steel bar of square cross-section carried by a couple of sets of rollers located on ball bearings. The front end of the bar was provided with a clamp for holding the electrode. Steel wires of 0·4063mm (0·016in) diameter fixed to the back end of the bar sustained the load, applied by weights to the bar to drive the electrode forward. The electrode had its front end fed through a guide somewhat greater in diameter than the electrode itself, and was thus firmly gripped so that throughout the test it maintained the proper position in regard to the lap joint. The complete feed mechanism was rotatable in the horizontal direction and vertically also, so that the electrode would be fed into the joint at whatever angle was desired.

The transformer used in the tests was of 500A AC with an open circuit voltage of 78. Current flow for each testpiece was metered through a relay in the welding circuit, and in each testpiece measurement of the weld length was made. Since the time and length of weld were established, a precise speed value for every testpiece was possible, and the testpieces repeated the state of the metal in production.

Larger diameter electrodes were later used, and in these instances the lap joints were 1,524mm (60in) long as compared to 1,219mm (48in). When three different electrodes for this type of operation were tested, considerable differences in performance were recorded, and could not be attributed to variations in the operator's skill or habits.

In general, electrodes should be tested in both the laboratory and service. Numerous cases are on record in which wide discrepancies have been found between laboratory tests and those given by the electrodes under field conditions.

Chapter 9

Of the many physical tests applied to tool steels in the past, few are now believed to have great practical value. However, some tests used are indicated below.

TESTS OF TOOL STEELS

The simplest and most rough and ready, once invariably used, but now relegated like many other practices to the small workshop, is the fracture test in which the length of steel to be tested is nicked on an anvil while still in the annealed condition, and an end broken off by a hammer blow. The fractured surface so exposed is then examined, care being taken that no contact between fractured surface and fingers occurs. This examination reveals to the trained observer any internal flaws, while the grain size enables him to determine the quality and properties with remarkable accuracy.

Decarburisation or soft skin is indicated by a variation between the sizes of the grains at the outer edges and the interior. Larger crystals with a brighter appearance at the outer edges show surface decarburisation, inevitable after forging and rolling to bar form. The purpose of the test is to determine the depth of this decarburisation.

Pipe, a cavity running down or close to the centre of the bar, is quickly revealed. Rupture caused by forging or rolling at too low a temperature is shown by a hole rather than a pipe, it being different in form. If it is difficult to distinguish between these two defects, further tests must be adopted.

Flakes show up as bright spots, or a short 'burst', in fractures taken longitudinally, but if the fracture does not run across the flakes, the bright spot will be somewhat differently tinted from the rest of the fractured surface, and will also have a conchoidal or shell-like form. Such flakes are harmful in water-quenched carbon tool steels, but not always in others. Flakes, however, must not be diagnosed when in fact the bright spots are produced by other causes such as the elongation of sections or rubbing in course of fracture.

Fibrous grain structure does not necessarily indicate faulty steel, especially in large pieces, but suggests the need for further inquiry.

Heat-treated bars are still on occasion examined by the fracture test, and the trained metallurgist can often detect faults more easily in this way. Sometimes a relatively thin disc is cut from the steel and fractured in the direction of mechanical working. This enables those defects present to be studied in their elongated state.

The fracture test also reveals the approximate carbon content of the steel and its uniformity over the entire fractured surface. While in most instances the higher the alloy content of the steel, the smaller its grains, grain size is also affected by general composition, properties, heat-treatment and quenching time at the hardening temperature, as well as the actual quenching medium. A fracture can also show to some degree how far a steel has been correctly heat-treated, but only in certain instances, and additional tests are usually required.

Some establishments used to provide a set of fractures graded according to grain size, successively numbered by size, and used as standards for comparison with other fractures.

Hardness and hot hardness tests, as well as the microscope and the dilatometer, are applied to tool steels, but these have been earlier described. Space must be found, however, for two other tests.

Macro-etching or macroscopy is applied to all types of steel and reveals the uniformity of the macrostructure.

The testpieces are polished and deep-etched with a suitable acid, having first been planed, ground or hand-filed, or *linished* by a flat revolving cloth belt whose surface is impregnated with some type of fine abrasive. Local heating must be prevented. After polishing has been completed, the surface is cleaned and deep-etched with 10% nitric acid or a copper reagent. Washing with hot water, cleaning and drying, are followed by a final rub with the finest emery paper, to show up the structure more clearly. The surface is then photographed or inked, and the impression transferred to a sheet of art paper.

Alternatively bromide paper is steeped in dilute sulphuric acid, laid on the surface and rolled with a rubber roller. During the rolling chemical reactions take place, after which the paper is stripped off, washed, and 'fixed' in a hypo solution. The result is an impression of the regions in which sulphides occur in the steel.

These tests show how the metal has flowed during mechanical working; where the various impurities lie; and where segregates are to be found. Impurities are often shown up in welded or soldered joints.

Abrasive tests are occasionally adopted to discover how far a tool steel will withstand wear. The loss of weight/unit surface area is accurately measured by moving the testpiece over a standard hardened and finished machined surface, with which it maintains contact for a predetermined number of strokes under a definite pressure, simulating as far as possible the working conditions. The loss is then compared with that of identically tested specimens of other steels of different composition.

THE SPARK TEST

The spark test is not only applicable to tool steels, but to many other types of ferrous metals. It is a simple and convenient method of sorting tool steels that have been mixed up by accident or negligence. The metal is ground on an abrasive wheel revolving at high speed, and the form and colour of the sparks given off are studied. The test is not a substitute for chemical analysis, but is useful with steels of different chemical compositions.

The steels to be examined are individually touched against the wheel, and the characteristic spark is given off. The test is widely applicable and can be employed for tool steels at any stage of production or use, being carried out on the bar itself. Steels can be rescued from the scrap heap when it would not be economical to test or analyse the entire batch in more elaborate ways.

The action of the grinding wheel sends minute fragments of white hot steel flying off at high speed, their passage through the air being readily discerned, particularly against a dark background. The path of the spark is termed the 'carrier line'. Low carbon steels with about 0·2% carbon give off sparks as shown in Fig 17 and 18% tungsten high speed steels as shown in Fig 18.

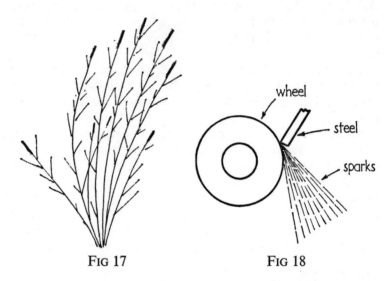

Fig 17   Fig 18

All steels contain some carbon, which at the high temperature of passage combines with oxygen in the air to form carbon dioxide as a product of combustion. The change from solid to gas increases the volume, the increase being resisted by the plastic steel particle. An internal pressure is therefore developed, ending in a bursting of the particle. The higher the carbon content, the greater the number of bursts.

Spark inspection demands skill and experience so that comparatively small variations in the spark stream can be detected and classified. A spark cabinet is used, of wood or other suitable material, measuring about 1,016mm (40in) long by 0·9144m (3ft) wide by 0·9144m (3ft) deep. This is mounted on a support, which elevates it to a height convenient for the operator, and its interior walls are coloured dead black. Electrical illumination is necessary, but direct light must be kept from the observer's eye and out of the cabinet.

THE SPARK TEST

The portable grinding machine uses an abrasive wheel at a minimum peripheral speed of 1,067m/min. The wheel is not of large diameter unless heavy grinding is necessary. It should be coarse and hard, weigh about 3·175kg (7lb), measure about 31·75 × 9·523 × 6·35mm (1·25 × 0·375 × 0·25in), and be placed parallel to the cabinet front so that the sparks it produces run at an angle of 90° to the observer's line of sight. Goggles should be used for eye protection.

The stream of sparks is horizontal and about 300mm long, so that precision in controlling pressure without reduction of the effective wheel speed is required. The wheel must not run too fast, as an inexperienced observer might then take the steel to have too high a carbon content.

Two quite different carbon steels, say 0·2% and 1·0% carbon, are tested, each piece being applied to the same spot on the wheel. The eyes are focussed on a spot about one third the distance from the tail end of the spark stream, and only those sparks crossing the line of sight are watched. Soon the form and appearance of every type of spark will become familiar as they are memorised. With experience one can determine the right amount of pressure.

The spark stream is divisible into the wheel sparks; that portion of the stream closest to the wheel; the middle portion; and the tail, farthest from the wheel. The observer studies the carrier lines, that is, the white-hot, fiery streaks marking the course of the incandescent particles, the spark 'bursts', and the typical sparks. The carrier lines differ in length, width, tint and number. The bursts differ in intensity, size, number, shape and distance from the wheel or from the carrier line ends.

The number and intensity of the bursts show the carbon content. Most alloying elements affect the bursts, carrier lines or spark form, speeding up or slowing down the carbon spark or darkening the carrier lines.

Silicon eliminates the characteristic carbon burst to a large extent. Tungsten high speed steels (18%) show a spark roughly as in Fig 18, reddish in hue, with a large number of red lines. Molybdenum in high speed steel does not wholly damp out the carbon bursts, but produces an orange-coloured spear tip at the end of each carrier line, but separate from it. They are always there whatever the length of the carrier lines, but are not easily detected if the carbon content of the steel is high. If the surface is slightly decarburised, however, the tip shows up more clearly.

Testpieces should be cut from the bars and be about 25·4mm diameter × 76·2mm long (1in diameter × 3in) long. The observer should have good eyesight and powers of concentration, be carefully trained and given regular practice. All spark testing should be properly supervised.

Lighting conditions include a dark background with the observer in shadow and no direct light. The air should be still to prevent strong currents from giving the terminal sparks a nook, so confusing the observer. The steel should make contact with the wheel at the bar end, not at the side, so that the spark stream does not come from a decarburised area. Forgings should be ground at the point where the tong grip was removed.

TESTING FILES

The testing of files begins at the factory with a number of simple ocular and aural tests applied to every file. It is first tested for straightness to ensure that it has not warped in the brine bath hardening process. It is also rung on a steel block to ensure soundness and freedom from cracks. The file teeth are examined in a good light to see that none are broken or burred, and finally a standard high carbon steel test bar is rubbed up and down the toothed file blade to test its 'bite' and whether any area is soft.

Concavity of the file surface is bad because the teeth in the concave area will not make proper contact with the work. Files, even those officially termed 'flat', should be slightly convex. Half-round, round and crossing files in blank form are tested by inserting them into holes of the correct form drilled beforehand in a steel plate. Three-square file blanks are tested for shape with a 60° thread or centre gauge. Square and flat file blanks have their corners tested by a thin steel square. Surface finish of blanks is also inspected and grooved or badly annealed material rejected.

The empirical testing of files has, of course, no scientific value, since it is affected by far too many variables, such as the degree

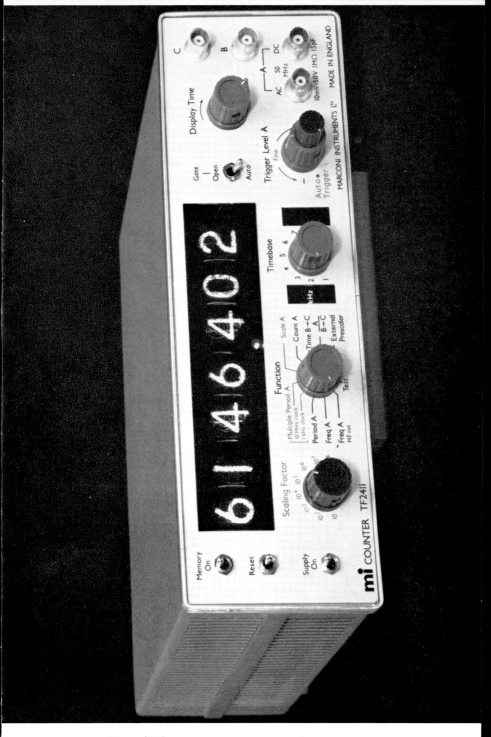

Page 167 Marconi Direct Frequency Counter for measurement up to 50MHz, with 0·1μsec timing resolution and auto-trigger

Page 168 Creep and Fatigue Testing Machine

of pressure, the kind of filing, the form of the file, the conditions of filing, and the physical capacity of the operator.

This was recognised in 1905 by Edward G. Herbert, who invented a file testing machine which automatically tests files under standard conditions, the results obtained being given in the form of diagrams produced by the machine itself.

The file is held between a pair of headstocks mounted on a reciprocating worktable, and has a stroke variable from 0 to 152·4mm (6in). The left-hand headstock has a handwheel and screw for adjustment of the file, its working face being parallel to the direction of motion. The file is reciprocated against the end of a test bar, supported on rollers and pressed against the file by a weight and chain giving a constant pressure. On the return stroke the bar is withdrawn to prevent rubbing.

The bar is reduced to filings and the file then ceases to cut. The diagram is produced on a sheet of section paper wrapped about a cylindrical drum, which is geared to rotate slightly after each file stroke, 25·4mm (1in) of the periphery representing 10,000 strokes. A pencil, connected by a fine chain to the test bar, is moved longitudinally across the drum as the bar is filed away. The compound motion of the drum and pencil gives a curve showing the action of the file at every moment of the test.

Tests can be made on bars of cast iron, steel, brass or other metals, and particularly on the metals most likely to be filed in service. The standard test bar for small files is 25·4 × 12·7mm (1 × ½in), one other standard bar being used for the larger files.

Stroke length and pressure of the file are varied to suit the different file lengths and the purpose of the test, the number of strokes/min remaining constant. The tests must correspond in number of teeth/cm on the files, size and metal of the test bar, length of stroke, number of strokes/min, and weight pressing the bar against the file.

Cutting rate at any period is given by the inclination of the corresponding portion of the curve. The nominal rate measures the sharpness of the file when new, irrespective of quality or temper of the steel. It is given by a tangent A (see Fig 19) cutting the line of 10,000 strokes at 96·53mm (3·8in), that is, the *nominal rate*. The *mean rate* is given by a line approximating to the

L

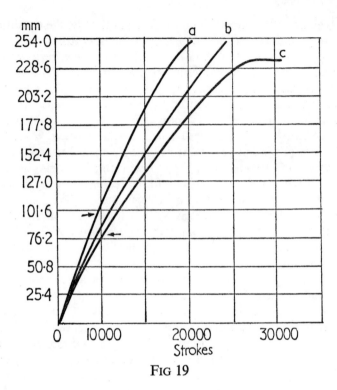

FIG 19

general slope of the curve, and cuts the lines of 10,000 at 76·2mm (3·0in).

The cost in files, wages and establishment charge incurred in filing away 100cm³ of metal with a specific make of file is the best test of its efficiency, and is given in new pence by the equation $C = \dfrac{100 P}{I} + \dfrac{380}{R}$, C being the cost of filing 100cm³, P the net price in new pence of each file, and I the total cm³ filed away by both sides of the file. R is the mean rate of cutting (cm/10,000 strokes).

Where the testing machine cannot be employed or it is not desired to have the files tested by an outside testing establishment, the only means of evaluating files is to try them out in service.

Frémont invented a special machine for testing files owing to his dissatisfaction with the Herbert test. He used a motor-driven machine at a normal speed of 60 strokes/min. A recording apparatus traced the diagram of work done, the track of the file and the thrust being used as ordinates. The file was held horizontally, and the machine was arranged for calibrating the spring by bringing the dynamometric handle into a vertical position by rotation, the paper of the diagram being fixed on a second board. This calibration was effected under graduated loads up to a weight of 200kg, corresponding to a maximum thrust under 100kg file pressure.

Professor W. Ripper of Sheffield University modified the Herbert machine, but the inventor did not accept his modifications, and they were never widely adopted. De Fried marketed numerous German copies or models based on the Herbert machine, he having bought the German patent, payment for which he completed a month before the start of the war of 1914–18. All these machines basically represented or imitated the forward and backward movement of the hand in filing. Buxbaum, however, developed a machine in which the file was stationary and the work moved, but not reciprocally, having an uninterrupted turning motion with no back or return stroke. The work was thus in continuous contact with the file teeth.

In this way a curved rather than a straight path was described, and by combining an eccentric movement with the turning movement a close approximation to the filing process was achieved. A table gave the different types of files and the corresponding test times required. After about 5min the file was taken from the machine for comparison with a master file of the same type, either by observation of the dulled teeth, or by measuring the decreased thickness of the file. Also an ammeter reading could follow the progressive dulling of the teeth, so that a decline in electric current strength could be used as a measurement of the file wear.

SAW TENSION TEST

Tension is the degree of rigidity given to the cutting edge of a saw to keep it straight in the cut. In general circular wood saws are tensioned to suit a working speed of from 48·25 to 50·79m/s (9,500 to 10,000ft/min), according to the working conditions and shop practice. For metal-cutting circular saws it is from about 60·95 to 91·43m/s (12,000 to 18,000ft/min), but recent developments in saw design and manufacture have enabled even higher speeds to be used.

The tension given to a saw is deliberate and necessary despite being a degree of strain. It is the force exerted to ensure tautness and rigidity during cutting. It is particularly important in bandsaws, which consist of a continuous ribbon of metal toothed on one or both edges, running over vertical pulleys and driven like a belt. The first step in tensioning a saw is to place the saw on a saw bench and try it with a straight edge or parallel strip used in checking the linear accuracy of work. The saw is examined inch by inch to ensure that it has no small humps, the saw being maintained perfectly level. After the humps have been flattened out by either hammering lightly and closely with a special hammer, or rolling, or both combined, a tension gauge is applied to determine where tension is inadequate.

The user elevates the saw, with the straight edge firmly across. If the saw centre falls away from the straight edge, the tension is correct. If the centre of the saw is too tight or insufficiently tensioned, it will cleave to the straight edge and will need no hammering or rolling. A second gauge is then applied, having an edge conformable to the right amount of tension for the particular saw. If a gap between gauge edge and saw is observed, the saw has excessive tension. Close fitting of blade to gauge signifies correct tension. If the gauge rocks on the saw, tension is inadequate. It is a common practice to turn the saw over from time to time to prevent the band from becoming dished or concave.

Excessive tension is rolled out. Inadequate tension requires the areas concerned to be rolled or hammered with a round-faced hammer along both saw edges, uniformly on each side. Metal has to be transferred in this way from edges to centre or from centre to edges, as may be necessary.

Rolling alone does not give tensions of proper uniformity, hammering is also needed, but must be uniform on both sides of the band. The hammer has to be carefully used, the blows not being too heavy and the tension and straight edge gauges being freely applied. The hammering should not be closer than 6·35mm ($\frac{1}{4}$in) to the saw edge. The hammer face must not be too sharply angled, as the notches it makes in the blade will cause cracks.

Tension is adapted in extent and location to the pulley face and wheel speed. Saws from 1,524 to 3,048mm (6 to 12in) wide fall from the level when tested to an extent ranging from 0·3969 to 0·7938mm (1/64 to 1/32in).

Twists in a bandsaw are usually caused by accidental contact with metal (in a wood-cutting saw) or by running in or out of the cut. To test for these, the saw is laid on the ground. If in correct condition it will stand up straight, but if it slopes in one or other direction, it is twisted. The length of twist is determined by rolling the saw along the floor until the whole length has been inspected. If it slopes at one point and is straight at another, the twist is short and affects only a small part of the band. If it slopes in all positions, the twist runs through the entire blade. A long-face twist is that seen when the saw slopes to the right. A cross-face twist makes the saw slope to the left. The double twist is one in which both left and right slopes are seen in the same saw.

For circular wood saws, the tension test is carried out in large timber mills and joinery shops by machines capable of testing the saw to a few thousandths of an inch for running truth. They also show the size of the bulges and the blue spots that indicate bulges. Tensioning today with all saws is still largely a matter of trial and error. Frequent testing is essential.

TESTING RIVETS
The shanks of rivets are usually tested by subjecting them to the normal bend test. A head is then formed on the testpiece by riveting over on a hard plate of steel, and if no cracks appear, this is an indication that the rivet is sufficiently ductile.

TESTING SINTERED CARBIDES
The increasing use of tungsten, tantalum and other metal carbides for cutting tools, etc, has rendered tests necessary for various reasons. One test is designed to measure the transverse

rupture strength so as to determine what angles of the cutting edge provide maximum tool support. This test is better than the ordinary tensile test for carbides because the testpiece is less expensive, a small tensile testpiece involves difficulties of alignment, and the tensile test is not suitable for brittle materials.

The testpiece is either rectangular, or round, and of beam type, the load being applied until the material ruptures. The value given is in kg/mm^2, using the ordinary beam formula, but the data do not represent strictly correct values owing to the great magnitude of the stresses concerned. Moreover, at large deflections, the effect of the bending load is to produce a variation from the correct stress values not accounted for by the ordinary beam formula.

The test is, however, a serviceable and effective means of grading the various materials according to their transverse strength, and the tests can, moreover, be repeated.

There is a tendency today to standardise the testpiece at 5·08 × 6·35mm (0·200 × 0·250in) with a distance between centres of 14·03mm (9/16in). Recommended procedures have been suggested and agreed upon for compressive strength, elastic modulus, thermal expansion, electrical resistivity, and Poisson's ratio, (the ratio of transverse contraction/unit dimension of a bar of uniform cross-section to its elongation/unit length when subjected to a tensile stress). Metallographic tests are also being standardised, as are apparent porosity, and density.

TESTING COMPOSITE METALS

These involve the testing of different but united metals. No one method can be applied to all, so that only general indications can be given concerning the best methods. For clad metals, that is, those in which two sheets of different metals are united by a bonding process, the bend test is often used, and consists of bending the sheet, in one case with the cladding part outwards and in the other with the cladding inwards. Each sheet is then bent a second time in the reverse direction, so that any flaw shows up at the second bend if not at the first.

The tensile test is useful for revealing the strength of the bond or weld, and this test is more stringent and trustworthy than the bend. The testpieces are usually flat, but if round testpieces are used, allowance has to be made for the effect of machining in altering the proportion of inserted to backing material.

For hardness tests the thickness of the testpiece should be 4·762mm (3/16in) in or above, and they can be tested in the Rockwell machine or the scleroscope, but specimens less than 1·587mm (0·0625in) thick are best tested by scleroscope. The more trustworthy method is to machine, grind or file off all the steel save where the hardness is to be determined.

Cupping tests on sheet metal of composite type involve drawing the metal so that the corrosion-resistant metal is on the same side as in the final piece or sheet. It is better that the insert material should be on the inner side of the cup.

It is sometimes required to determine the position and dimensions of the insert. One method of ascertaining this is by immersion in a warm bath of about 50% nitric acid for half a minute. The insert will then show up, and the solution should then be removed by warm water. The next step is to apply some non-attacking protective covering, such as oil, over the etched area to prevent further attack.

Another method is often employed for carbon tool steel or other plain carbon steels, namely to take advantage of the phenomenon of temper colours. The clad sheet is heated until it turns a light blue or pale yellow on the steel insert, which is revealed quite clearly. The test takes longer than the etching test in that the insert is not so rapidly disclosed.

TESTING METAL PLATINGS

Electroplated metal coatings are usually covered by specifications which demand a minimum thickness. This is determined in the orthodox manner by removal of the coating from a specific area. The test may be one of weight loss or of chemical analysis. The thickness can be measured in various ways, such as (a) microscopic inspection of a chosen cross-section; (b) cutting through the plating with an abrasive wheel of specific radius; (c) testing the rate at which certain chemicals react, such as those used for zinc, chromium and cadmium coatings; (d) magnetic attraction; and (e) microscopy, which is the most trustworthy save where the platings are exceptionally thin.

The minimum thickness depends on the period of time and current density used in the plating operation. Another important factor is the distribution of the plating over the surface, a variable determined by the solution power, the form of the plated piece and its location in relation to racks and anodes.

Platings are usually dense, but porosity may occur, defeating the object of plating. Some of the finer metals such as gold, silver and platinum, when used as coatings on steel, may be tested for porosity by salt spray or by the ferroxyl indicator test in which a water solution of agar agar 10g/litre, sodium chloride 10g/litre, and a few drops of phenolphthalein, form a solution in which the testpiece is immersed. The phenolphthalein reacts with hydroxyl ions to give a pink coloration showing the degree of alkalinity at the cathode portions of the metallic surface, the ferrocyanide reacting with the iron ions to produce a blue coloration where the iron is being dissolved at the anode.

The ability of the plating to adhere to the base metal under stresses of various kinds is not quantitatively determinable. Various deformation tests have to be used to show how the coating behaves under the type of stress or stresses to which it will be subjected.

Hardness cannot be correlated with performance in use owing to the wide variations shown according to the plating procedure. Abrasion resistance is reasonably well indicated by the scratch hardness test.

The brightness of the plated surface has been tested by determining the ratio of the luminous flux regularly reflected from a surface to the total flux falling on the surface, that is when the reflection angle equals the angle of incidence. In the standard test of this type the reflection angle is taken as 45°.

TESTS FOR SOLDER

The value of a solder lies not so much in the metal itself as in the strength of the joint made with it. For this reason there is little point in testing the tensile strength, hardness, ductility and other characteristics of the metal. Instead, the joint strength is tested, usually by established procedures such as tensile testing, shear tests, spreading and wetting tests, and the degree to which the molten solder climbs by capillary action between the pieces to be soldered. These last three are highly specialised tests, the details of which are best obtained from the makers of the particular solder used.

TESTING TINPLATE

The most important test to be applied to tinplate is the corrosion test, since these plates are still extensively used for canning foods or as tops for bottles and cans. The accelerated corrosion test is often employed, the filled can or other receptacle being maintained at adverse temperatures to simulate storage in tropical or subtropical climates. The length of time that the contents take to spoil is measured. Another and quicker test still is to measure for a specific volume of hydrogen to be given off after the plate has been soaked in a bath of warm dilute hydrochloric acid. For this test the conditions must be standardised.

STRESS ANALYSIS

When a new product is being designed, embodying components of intricate form, the designer needs to know in advance, as far as is possible, the way in which the stresses encountered in service will be distributed, their type and severity. This means that he must devote his time to a number of activities known as 'stress analysis'. For this he will require highly intelligent and trained operators and assistants, as well as the ability to call upon an extensive range of appliances and procedures. Among his needs will be included such items as electric resistance strain gauges capable of being firmly attached to the surface of the component to record the specific changes in resistance according to the surface strains. He will also need capacitance and induction gauges, extensometers to work on short gauge lengths, and on occasion he may need to use some type of brittle coating or lacquer applied to the metal surface by spray. This sets, and when the component is stressed, a pattern of cracks results, the cracks running at right angles to the greatest main stress direction. The lacquer is in effect a strain gauge whose length is exceptionally short.

The simple lacquer method is commonly termed the *Stresscoat method*, 'Stresscoat' being the name given to the lacquer originally developed in the United States. It is available in a range of qualities conformable to the atmospheric conditions of the tests. The first crack is observed at a strain value of 0·0005 to 0·001mm/mm of thickness.

The lacquer itself is tested for sensitivity by the formula $S = E \times e$, S being the sensitivity of the lacquer and e the strain value, E being the elastic modulus. S therefore indicates the precise stress at which the first crack will occur. Later cracks obtained by increasing the load in specific amounts are recorded, should progressive values of stress be needed, and when the entire crack pattern is visible, it shows the distribution of tensile stresses over the component.

This brittle lacquer method is also valuable in the testing of parts in motion, such as rotors running at high speed, but here the object is to make comparison between alternative designs, the data obtained being not so much quantitative as qualitative.

The disadvantage of the lacquer is that unless strict regulation of the temperature and moisture content of the atmosphere can be achieved, the results will not be entirely trustworthy. Such a degree of precise control is not always possible. Another draw-back is that most of these lacquers produce a spray of poisonous type, so that great precautions are necessary when spraying.

The best and most satisfactory means of determining in advance the distributions of stress within a component when under load is by *photo stress*. In this, a thin film of a special material is cemented on to the metal surface in advance of loading. Polarised light is passed through the film, which may be of nitrocellulose or some material having the necessary optical properties. It is then possible to calculate the magnitude and direction of stresses at all points in the component. New plastic materials have considerably increased the scope and value of this test, but these demand study by an expert in the technique. A reflection technique is commonly used to determine the stress condition of any particular surface. Sometimes a complete model of the component is made and lacquered for the test.

The *frozen stress* method is also used, the model of the component being loaded in exactly the same manner as the part will be loaded in use. The model is then sliced, and the slices studied in a polariscope, comparative measurements of retardation being taken of the permanent bi-refringence the operation induces. This bi-refringence is the property of an anisotropic crystal when studied under cross nicols (prisms). It produces typical tints constituting a measurement of the difference between the maximum and minimum values of the crystal's refractive indices.

The measurements so taken enable a quantitative estimate of the stress in the component to be obtained. Those parts of a component of maximum importance to the designer are the concentrations of stress almost always found on the external surface at locations where a principal radius of curvature is small.

Other three-dimensional methods of ascertaining stress distributions are scattered light or sandwich techniques, which are highly specialised and for a description of which space is not sufficient.

The analysis of stress by photo-elastic methods is based on a discovery more than 100 years old, when it was shown that sheets of glass subjected to stress revealed coloured patterns under polarised light. Polarised light divides into two parts or beams which travel in different planes. When a stress concentration occurs, it changes the velocity of the two beams so that they interfere with each other optically, forming patterns whose lines are nearest to each other where the stress concentrations are at their maximum.

Consequently, stress analysis using this method allows the designer to determine in advance the entire stress distribution pattern, and also shows him the stresses in other zones he could not otherwise see. The usual practice is to use the three-dimensional method despite its slower technique, but if the model of the component is of plastic material, this is unimportant. The model so made records the same stresses exactly as a metal model, since stress distribution is governed wholly by the geometrical shape of the part.

ELECTRICAL STRAIN GAUGES

These are now extensively used to provide a quantitative analysis of stress, and can be used in both static and dynamic testing. They are adaptable to rotating parts and the accurate measurement of centrifugal, torsional and vibratory stresses at varying speeds. Originally they were of carbon type, but today a nickel chromium wire in diameter not greater than 0·0254 × 634·9mm (0·001 × 25in) long is used, in spiral form so flattened that the turns form a zigzag about 25·4mm (1in) in length × 15·87mm (0·625in) wide, but alternative sizes are obtainable. The ends of the wire have welded or soldered leads, and the assembly is cemented to a strip of paper itself similarly attached to the part to be tested.

If the metallic object to which the wire is cemented undergoes extension or compression, the wire changes in length and electrical resistance when an electric current is passed through it. This variation in resistance is measured by a Wheatstone bridge consisting of two parallel resistance branches, each branch containing two resistances in series. Attached to a portion of metal resembling the component being tested is an inactive gauge which is not under stress. This is inserted in the circuit to even out any resistance changes caused by changes of temperature. The inactive gauge is of identical resistance to the live gauge (usually about 200Ω) and of an accuracy of 0·1Ω.

When rotating parts are being tested, the gauge wires are linked up with the instruments by way of slip rings and carbon bushes so that stresses may be registered.

Chapter 10

LONG TERM CREEP RUPTURE TESTS
In the United States superalloys are being tested by short-term tests of less than 1,000h to show their long-term creep rupture strength. The superalloys tested, designed for gas and steam turbines, have a composition (nominal) of 0·05 C, 0·10 Mn, 0·10 Si, 12·60 Cr, 43·00 Ni, 6.00 Mo, 4·00 Co, 3·00 Ti, 1·25 Al, 0·01 B, and about 28 Fe, %. They are made by a combination of vacuum induction and consumable vacuum remelting. Tests previously designed for an alloy capable of withstanding a temperature range from 548 to 820° C (1,000 to 1,500° F) lasted more than 4 years, and showed the trustworthiness of the data obtained from the short-term tests, as well as the great stability of the metal.

THE 'FASTRESS' ANALYSER
An automatic stress analyser capable of measuring the residual stress of a metal subjected to tensile or compressive stresses has been developed, and measures these in 20s. It is claimed that it can be employed for monitoring heat treatment, cold forming, welding, shot peening, grinding and other production processes. The analyser contains two X-ray sources placed 60° apart.

GEAR TESTING
A new tester has been devised by a British company to reduce an entire range of expensive and slow tests for gears to a single rapid comprehensive measurement. To measure gear transmission errors, two component gears or one meshed with a master are mounted on two precision spindles adjusted to mesh at their designed centre distance. One shaft is then motor driven and drives the second shaft by way of the gears under test. The relative motion of the driving shaft to the driven is then continuously

measured through one revolution and a single interrelated graph records the final quality of gear transmission.

The motions of the two shafts are compared as regards phase and the consequent trace shows the rise or fall of angular speed. These fluctuations normally appear in cyclic form directly related to the number of teeth in mesh, and can be attributed to the different gear errors. The most advanced gears can, it is claimed, be measured, and major form errors, errors of pitch and eccentricity, are all readily ascertained from the graph. The machine has been successfully tested under the auspices of the Ministry of Technology.

ELECTRON FRACTOGRAPHY

Some errors in manufacture or processing start fractures too minute to be seen by microscopy. Only recently has it been found possible to detect the origin of certain fractures by the aid of the electron microscope, using extremely high magnifications. Flaws that can be detected in this way include grinding and quenching cracks, hydrogen flakes, hydrogen embrittlement fractures, flow-throughs, cold shuts, seams, non-metallic inclusions and porosity. The fractographs are prepared by 2-stage plastic carbon technique and shadowed with chromium at 45°.

The practical value of the technique is shown by numerous observed cases. In one, for example, ground maintenance personnel for aircraft detected a leak of hydraulic fluid from two small cracks in a main landing gear cylinder. This was found to be caused by the failure of a steel component having a tensile strength of 1,791 to 1,895 MN/m^2 (116 to 120 tons/in^2). The failure began on the internal diameter of the cylinder, numerous cracks being found under the chromium plating. The electronic examination revealed that the cracks were principally intergranular with hairline indications caused by hydrogen embrittlement occurring during or after overhaul.

CLOSED-LOOP ELECTROHYDRAULIC TESTS

Service simulation tests are becoming increasingly popular, as with such parts as landing gear needing true dynamic loading tests, and in the production of aircraft, automobiles, lorries, marine craft, tractors, etc. The closed-loop system is claimed to give ultimate control of load, strain and stroke for fatigue testing either parts or complete full-scale prototypes. The systems can duplicate any wave form, including sinusoidal forms, actual service recordings and random noise.

RESONANT FATIGUE TESTING

Recently the world's largest high speed fatigue test system has been installed at a large American steelworks. The system is said to give up to 453·6Mgf (1 million lb) of tensile load at frequencies from 20 to 50Hz (cycles/sec) with up to 907·2Mgf (2 million lb force) for frequencies up to 2Hz. The system is used for testing huge fasteners and bolts for aircraft and diesel engines.

COMPUTERISED TESTING

Also in the United States a constant resolved shear system has been developed to maintain a constant stress across a testpiece as its neck diminishes. The machine is controlled by an analogue computer, and will test specimens under loads ranging from 1·26 to 2,268kg (3 to 5,000lb) in vacuum down to 10^{-8} Torr at temperatures up to 1,090° C (2,000° F). The testpiece train is said to hold bending to 3% or less at a 7·693kg (17lb) load. The system is being employed for basic research, including studies of individual grains.

HYDRAULIC TESTING MACHINE

The modern testing machine for hydraulic testing has flexibility and maximum ease of operation. 0·5% of dial reading and 0·1% of range capacity are demanded for accuracy over the entire system. The dial is masked, colour coded, has up to 1,200 widely spaced graduations for more precise load reading, and is about 711·2mm (28in) diameter. The dial illumination is neon, glare-free, and provides for 3 or more range capacities with a touch of a switch. It may also have a patented auxiliary pressure system assuring friction-free loading, and maintaining accurate off-centre loading.

In addition it may have built in zero compensators, range capacities from 136 to $2·268 \times 10^3$Mgf (300 to 5,000,000lb), and a complete range of adaptations to match the physical testing needs of the user.

THE MODERN METALLOGRAPH

The tendency in designing modern metallographic cameras is to build into the machine as many as possible of the basic requirements necessary. Such a machine has, for example, flat field achromat objective as a standard (planachromats being optional, $1\cdot3 \times$ to $100\times$), a direct-reading exposure meter on the control panel, which reads colour temperature for ultra-accuracy in colour work. It has ground glass and eyepiece for viewing and focussing, and a sealed-in, dustproof shutter and flat field photo eyepiece turret. Direct-fitting 35mm and Polaroid camera backs are interchangeable in seconds, and there is a mechanical stage. Options include tungsten halogen high intensity illuminators, oblique and trans-illuminators, grain-size and measuring eyepieces. Visual observation at magnifications from $5 \times$ to $2,000 \times$ are possible. Additional accessories include special devices for interference microscopy, plating thickness determination, high temperature studies, etc. Metallographs are also available with xenon illumination, filters, micrometer reticule, etc.

THE HYDRAULIC FATIGUE SYSTEM

This is a versatile new testing machine based on experience which enables the user to carry out all bending, direct stress and multiplying fixtures. Frequency, load and amplitude vary under fast electronic control throughout the test, it is claimed. The system offers $+$ and $-$ 3·175mm (0·1250in) amplitude at 30Hz and up to 152·4mm (6in) total amplitude at reduced frequencies.

In the fatigue machine, a programmed electronic function generator controls a 2-stage servo valve that regulates load. An electronic load cell measures load, and a force loop allows the operator to control load manually or with a programmed input. If the test programme calls for independent amplitude control, a position loop can be added to give manual or programmed control of amplitude as well as load. Moreover, a strain loop can also be added to the system.

To make setting up easier, the position of the loading platen can be adjusted from the loading frame. The function generator produces square, triangular and positive or negative slope wave forms, as well as the standard sine wave. The load up to 4·536Mgf (10,000lbf) can be applied gradually, or can be released suddenly or applied in a single direction or two directions. Frequencies are within a range of 5 to 200Hz.

Standard fixtures multiply the versatility of the system, while special fixtures widen the applications.

THE NEUTRON RADIOGRAPH

Many assemblies of metal case type are impossible to study by X-ray, such as explosives, pyrotechnical devices, composites, mechanical and electrical assemblies, and hydrogen at extremely low levels in metals. However, by neutron radiography it becomes possible to obtain excellent contrast and also details of an interior that cannot be seen in the X-ray radiograph.

Whereas X-rays interact with electrons orbiting atomically, so that elements with a large atomic number are difficult to inspect, neutron interaction is unaffected by atomic number. The range of absorption coefficients for neutrons is 0·03 to 90, which is much greater than for X-rays (0·13 to 4).

In making a neutron radiograph the piece is placed in a neutron beam, and some neutrons are absorbed, some pass through and some scatter. The image produced is recorded on ordinary X-ray film using an intermediate transfer screen of gadolinium, indium, etc. This absorbs the neutron image, which produces an ionising radiation image recorded on X-ray film. When developed in the normal manner, this reveals the internal details.

This system holds out many interesting possibilities, and readers requiring further details should apply to the Applications Engineer, Irradiation Processing Operation, Nuclear Energy Div, General Electric Co, Pleasanton, Calif, USA.

PORTABLE HARDNESS TESTER ON NEW PRINCIPLE

A British development is a new portable hardness tester working on a quite different principle from any previous instrument of its type. It is being used to test the quality of bolts in bridges and other *in situ* positions. The instrument measures the deceleration force applied to a piezo-electric crystal behind the probe when it strikes into the surface tested. It is claimed that consistently accurate readings are given, comparable to those of laboratory instruments. Measured in volts, the results are expressed in Equivalent Vickers Units, the range being 100 (low carbon steel) to 700 (hardened steel).

Batteries operating the instrument can be recharged from standard mains, and accuracy exceeds $\pm 3\%$ over battery voltage variations between 22 and 29V. An ambient temperature change of over 40° makes less than 1% difference to reading accuracies. 100,000 operations produce negligible wear. Simplicity of operation and availability of different forms of anvil are advantages when testing rods, bars, machined components, dies, etc. The machine has been developed by the Research Unit, Guest Keen and Nettlefold's Group Technological Centre, Wolverhampton.

NEW PORTABLE STRAIN INDICATOR

This machine weighs a mere 3·175kg (7lb) and is battery powered, but eliminates bridge unbalance, it is claimed, and equalises everything to initial zero, while bringing selectable internal bridge completion for full, half and quarter bridge operation. It is robust, low in cost and accurate to $\pm 0·1\%$. Other developments in this field include strain gauges with their own adhesives on the back, and new scanner and recording systems with calibration independent of bridge excitation.

TESTING NEW LORRY CABS

Simulated service tests are used by the Motor Truck Engineering Dept of the International Harvester Motor Truck Div, Ft Wayne, Ind, USA. The prototype cab assembly is installed on a chassis, and transducers such as load cells, strain gauges and accelerometers are located inside and around the cab to monitor electronically all stresses and strains. Each device is linked up with a suitable tape recorder.

As the lorry runs down the road on test load, data and vibration frequencies of pitch, roll and vertical bounce are recorded.

The vehicle is held at a speed as near as possible to that of normal operation. When the run is over and all data recorded, magnetic tapes 25·4mm (1in) wide are computer analysed. Histograms of input loads of the various transducers are provided, and a spectral analysis of each condition obtained to determine the frequency response characteristics. Should a component appear to be breaking down, the test is halted for suitable changes. The test has the advantage of giving in a few weeks results hitherto obtainable only after years of actual service.

NEW PENETRANTS FOR MAGNETIC TESTING
New penetrants of economical type have been developed and give, it is claimed, improved sensitivity, shorter processing cycles, more coverage/m^3, less evaporation and a greater contamination tolerance. Among these new types are water-washable high sensitivity fluorescent penetrant groups, aqueous remover systems, non-aquaeous remover systems, and broad spectrum sensitivity post-emulsification materials designed to locate the exceptionally fine laps or cracks, the medium size and the open defects, in a single test.

PORTABLE ULTRASONIC TESTING MACHINE
This machine, like the penetrants mentioned above, is the invention of the Magnaflux Corp, Chicago. It is battery-operated, and detects casting cracks, inclusions and variations in wall thickness of castings ranging from 381 to 762 × 1,270mm (15 to 30 × 50in) in one plane. Characteristic defects revealed include contraction voids and gasholes, which are clearly shown on a cathode ray tube and marked by an inspector for later analysis. As machining proceeds, the machine can be used to reveal hidden wall thickness in steam rings and valve bodies and to detect small subsurface voids.

ELECTRON MICROSCOPE
The modern electron microscope takes 6 testpieces at one time and embodies an electronic position indicator, with lighted 'nixie' tubes that display magnification from 125 × to 500,000 ×. A push-button gun lift allows the filament to be changed in

seconds every 150 to 200h. Printed identification numbers mark the films or plates and a hydraulic chair is used. Cameras, air locks, vacuum systems, are all automated.

IN-PROCESS CONTROL SYSTEM

An in-process control system for machine tools automatically corrects cutting or measurement errors and greatly improves accuracy and productivity of turning and boring machines, it is claimed. Once work has been set up on, for example, a lathe, pneumatic proximity gauges acting as sensing elements monitor the cutting, and immediately signal an error if there is any deviation from the workpiece. The signal operates a valve that allows the tool to be relocated. Tests have shown that quite small errors can be revealed in this way. The system was developed by the National Engineering Laboratory, East Kilbride, Glasgow.

NEW TYPE OF EXTENSOMETER

A new range of extensometers has been produced for direct measurement of strain and for use over a temperature range from -73 to $200°$ C (-100 to $400°$ F). They are stated to give exceptional economy and versatility for testing metals. Other new developments are a new transverse strain sensor for diametral evaluation of strain, a new high magnification calibration fixture allowing full calibration from 25·4mm (0 to 1in), and a number of grips including a high temperature pneumatic cord and yarn model.

TESTS OF STRESSES IN ALUMINIUM

The USSR have carried out some interesting tests in connection with repeated load application and the part played by microstresses during the relaxation of macrostresses. When stresses are relaxed at normal room temperature the lattice structure is deformed by slip, and becomes an irreversible plastic deformation. It was assumed that when the stresses were relaxed orientated microstresses set up might change the later behaviour of metals and their resistance to relaxation. It seems that orientated microstresses are actually generated on the initial load and revealed when later loads are applied.

Two groups of tests made involved (a) a test of relaxation in which the testpiece was loaded in the elastic zone, held for 200h at constant overall deformation, the load removed, and then held unloaded for a succession of similar cycles; (b) the second group of tests were largely the same, but no measurements were made of the deformation after the removal of load when the initial period terminated. To obtain measurements of the lattice constant and then the change in elastic deformation, the testpiece and the relaxation test block were mounted in a back-reflection X-ray camera.

The change in lattice constant corresponded to the elastic deformation of the testpiece. It was easily possible to compare the two sets of test results. The metal of the test was a 99·99% pure aluminium, vacuum furnace annealed for $1\frac{1}{2}$h at 200° C (390° F), with a grain size of about 0·05mm.

After the load is removed from a testpiece, the orientated microstresses become partly relaxed, as shown by a reaction, and if the load is repeatedly applied, the relaxation of microstresses occurs by reason of the effect of the orientated microstresses at a lower rate, so that the value of k_1 declines. K_1 is, of course, a constant.

HIGH TEMPERATURE FURNACE

A system of high temperature testing usually embodies a special furnace. Tensile, compression and flexure testing to 3,000° C (5,000° F) in high vacuum or controlled atmosphere is possible with a modern furnace, which has bracket mounting to fit all standard test machines of 558·8mm (22in) width clearance and 1,016mm (40in) between cross yoke load connections. Optical measurement is done through two sight windows provided with anti-fog diffusers. Optional accessories include high vacuum systems, optical dilatometers, vacuum extensometers, automatic temperature control and recording and programming.

THE MODERN BRINELL TESTING MACHINE

The latest machine of this type applies an initial load to seat the testpiece and zero the depth indication pointer. The machine is then automatically switched to a pre-set time for a pre-selected load. The application of the load is guaranteed to within $\pm 1\%$, and indentation depth is shown to the nearest 0·00254mm (0·0001in) in a front-mounted gauge.

The gauge has two pointers capable of being set at the high and low limits of hardness acceptability, so that the indicating system is one of 'go' and 'not go'. The machines work on compressed air with a minimum pressure of 448·1kN/m² (65lb/in²).

ELECTRONIC RESISTANCE WELD TESTING
A recent new machine of this type connects directly to the welding machine electrodes. As each weld is completed, voltage across the electrodes is measured, time, heat and pressure being analysed. The results, shown on a read-out dial, tell the operator if the weld is unacceptable. For calibration, the welder controls are set for the best possible weld, and the indicator needles adjusted for zero deflection. Upper and lower limits are then established by varying the input of heat until a 'hot' and 'cold' weld is produced, whereupon the limit indicators are set to suit.

LARGE SCALE NOTCH TOUGHNESS TESTS
The ability of a steel plate to withstand the propagation of cracks while subjected to high stresses at low temperatures is being studied in Australia for steel designed for use in a bridge. The plate material is a silicon-killed, aluminium-treated, carbon-manganese steel control-rolled to give good notch toughness. The test is designed to simulate adverse conditions, such as when a crack enters at high speed from an attached member, or a high speed crack generated within itself is caused by damage from an external source.

Unless the steel is tough, the crack will spread over the whole plate.

Plates 19·05mm (¾in) thick by about 254mm (10in) wide × 6·096m (20ft) long are welded into a double web beam, both flanges and webs being taken from the one plate. Small high carbon embrittled steel tabs are welded along the flange edges at intervals. Each tab has a saw-cut, extending only partly across the tabs and directed towards the beam to serve as the crack source. The crack is made by a bullet fired into an undersize hole lying diammetrically across the saw cut near but not quite at its tip.

The beam is loaded in 3-point bending, which enables the nominal stress near the tab to be varied with the load and the distance of the supports from the tab. For each test a short beam length about 457·2mm (18in) is cooled to temperature, measurement by thermocouples being made of the flange and web temperature. The bullet crack runs at high speed to sever the rest of the tab, crosses the weld and enters the beam.

Three patterns of behaviour are seen: (a) the crack is almost at once arrested in the beam; (b) it runs partly through the flange; (c) it fractures the flange completely, being arrested in the web.

The tests demonstrate that the steel, which contains 0·16 carbon, 1·33 manganese, 0·24 silicon, and 0·035 aluminium, %, could withstand fast running crack propagation even when subjected to high stresses at low temperatures. At the design stress of about 169·84MN/m^2 (11tonf/in^2), the steel arrests a fine running crack at $-40°$ C ($-40°$ F), and it is estimated that the steel will halt cracks at $-5°$ C ($-20°$ F) when under stresses greater than the yield stress.

AUTOMATIC NON-DESTRUCTIVE TESTING

The most important tendency in non-destructive testing for the automobile industry is an automatic in-line system that tests steel billets, for example, and combines ultrasonic and magnetic particles methods. On long billets, scanning speeds of up to 54·86m/min (180ft/min) are possible. The modern practice is to instal special equipment at strategic points in a production line to establish and control product quality by feeding details to the operator or directly into controls facilitating corrective measures.

For example, it is feasible to check steel almost continuously as it changes from a raw metal to a complete component and to an assembly. If the metal fails to meet specification, the faulty component is either discarded or returned for reclamation, while the information concerning the fault is made use of in altering process controls.

The machine, 19·2m (63ft) long by 7·619m (25ft) wide, deals

with 3 billets/min, either carbon or alloy steel, square, rectangular or round, 3·658 to 12·19m (12 to 40ft) long and 1,351·7 to 5,161·6mm² (2¼ to 8in) in cross-section. An 'unscrambler' feeds the billets into the ultrasonic test individually and the sound waves reveal internal flaws along centre lines perpendicular to two neighbouring sides of the billet. Two transducers of the machine oriented at 90° to each other are held in place by a floating carriage assembly following the camber or twist in the billet. A water column transfers sound from transducer to billet surface. Black light examination, after the residual wet method of fluorescent magnetic particle testing, then discloses invisible surface irregularities, heat cracks and forging laps.

INTERFERENCE MICROSCOPY
From West Germany comes a new differential interference microscope which is claimed to give surface examination a new dimension. The system is one of versatile phase and amplitude-contrast vastly improving images of surfaces not having large enough colour differences, refractive indexes or reflectivity for successful study by normal methods of reflected light.

The contrast is optically generated by interference of light waves traversing somewhat different optical paths. The differential interference produces an optically shadow-cast image, creating a 3-dimensional effect and bringing surface differences into high relief. It is easy to set up and operate, and has important applications where it is impossible or impracticable to etch or treat surfaces with chemical reagents. As examples may be mentioned the inspection before processing of polished semiconductor wafers for smoothness, stacking faults, holes or contamination, and the examination of unetched surfaces of electro- or cloth-polished metals and alloys.

A MODERN SERVO POTENTIOMETER
An American servo potentiometer has many interesting features, such as corrosion resistant platinum slidewire giving twice the resolution and length of life of former devices, it is claimed. Also incorporated are an input filter with full floating shield to withstand stray effects; a robust chassis casting for vibration-proof

servo drive train alignment; pneumatic proportional and 3 types of 3-mode electric proportional control units; and cascade zener diode network for constant voltage reference for measuring circuit despite fluctuations in line voltage or ambient temperature. A flexible drawbridge allows the chassis to be withdrawn in seconds, and a reversible terminal panel allows for front or rear power and process connections.

XERORADIOGRAPHY

This is a recent development of somewhat similar character to radiography. The plate or film is exposed to X-rays, which are differentially discharged at whatever point is impinged upon by the radiation. By this means an electrostatic image is obtained, rendered visible by plastic dust. Because the dust is electrostatically charged with a polarity opposite to that of the plate, the developing dust adheres to zones on the plate not discharged by radiation. For permanent record a photograph is then taken of the image, or the dust is removed on paper specially treated.

MEASURING NON-METALLIC INCLUSIONS

In counting non-metallic inclusions considerable error is possible, so that various automatic instruments have been or are being developed in which light optical or electron optical systems are used. The electron optical instrument and detection system is based on the principle that the number of back-scattered electrons is governed by the mean atomic number. A machine known as the 'Quantimet' has, however, been devised by the Metals Research Dept of the Melbourne Research Laboratories, Melbourne, Australia, in which the image from an ordinary optical microscope is projected on to a television camera screen, whose camera output passes into the detector unit. This is so set that it will measure total particle projected area, total particle number and length, particle size distribution and total particle chord size distribution, the data being printed out automatically.

The necessary movements on the particular field having been performed, the stage is automatically moved and the measuring procedure repeated. The speed is slightly higher than one field of view/s, but in that time feeds out results on 30 parameters, operational speed being limited by the recorder used. The optical microscope can, if desired, be replaced by an epidiascope attachment.

This instrument is claimed to be a general quantitative microscope and therefore a versatile laboratory instrument. It is currently being used for work on inclusions in ingots, and if employed with care and a knowledge of possible errors, is said to provide quantitative information concerning inclusion sizes and distributions not previously obtained. Its use in routine quality control has not yet been assessed.

THE FUTURE OF TESTING

The main stress in regard to the future of testing is being placed on safety and the trustworthiness of the product, so that more stringent testing and inspection will be demanded. Non-destructive testing especially in nucleonics is being developed intensively, and when combined with automation will greatly improve test results. For this reason attention should be particularly paid to developments in ultrasonic testing, X-ray spectroscopy and other appliances.

Dependence on manual skill is being increasingly eliminated, and in future automatic inspection and control will not only be necessary, but will be carried out at ever higher speeds as designers become more and more aware of the need for reliability in their products.

Defects will be more strictly measured and test data will be required in more detail, so that electronic processing by computer will be increasingly necessary. The testing of metals at high temperatures will also increase to enable faults to be revealed early during the various stages of manufacture.

The principal properties that will be demanded of products is their cleanliness and freedom from superficial flaws, their metallic soundness, their consistently good microstructural condition and their freedom as far as possible from residual stresses. It will be expected of an engineer that he shall forecast the length of service life that may be anticipated from his products especially under conditions of fatigue.

THE FUTURE OF TESTING 195

The depths and hardness of carburised and nitrided cases on steel, the composition of a steel melt, and the homogeneity of welds, will all be tested with ever greater precision. Fracture mechanics will become established by the aid of eddy current and ultrasonic tests adapted to processing by computers; and control of homogeneity of microstructure, elasticity and strength predictions will all be called for. Remote testing procedures including acoustic emission and acoustic spectrometry may become more usual as nuclear reactors need to know whether heat, pressure, stress and irradiation will enlarge cracks and fissures of microscopic or submicroscopic type into potentially perilous defects.

Remote quantitative tests may be used in conjunction with high-speed computers to indicate automatically the condition of a metal component in use, so that it can be removed and repaired or a new component be introduced in good time.

The scanning electron microscope, now of great and increasing power, will greatly promote fractography and render possible less stringent safety factors with no loss of reliability. The storing of test information in computerised banks is certain to be a feature of every testing division, and will be drawn upon by designers eager to choose the most effective and economical metal for every purpose.

Already on the horizon are laser beam techniques, scattered light procedures and infra-red devices, while a comprehensive numbering system for all metals and alloys is the subject of activity by at least two of the largest research organisations. Microprobe analyser units and stress wave emission are being investigated as means of inspecting metals for defects. The other potential developments space does not allow us to discuss here are so exciting that merely to enumerate them would stimulate the imagination of an H. G. Wells. The next decade will see startling advances.

Acknowledgements

Acknowledgement of information and or photographs is made to:
The Wilson Instrument Div, of Acco (American Chain and Cable Co), New York.
The Whiton Div, Terry Steam Turbine Co, New London, Conn.
The Magnaflux Corpn, 7322 West Lawrence Ave, Chicago, Ill.
The Tinius Olsen Testing Machine Co.
The Welding Products Div, Falstrom Co.
Astro Industries Inc.
Irradiation Processing Operation, Nuclear Energy Div, General Electric Co, Pleasanton, Cal.
Satect Systems Inc, Grove City, Pa, 16127.
International Harvester Motor Truck Div, MTS Systems Corpn, Ft Wayne, Ind.
Group Technological Centre, Guest, Keen and Nettlefold Ltd, Wolverhampton.
Automation Industries Inc, Materials Evaluation Group.
Ametek Testing Equipment Systems, Dept 50, Lansdale, Pa.
Olympus Corpn of America, Precision Instrument Div, 2 Nevada Drive, Hyde Park, NY 11040.
MTS Systems Corpn, Box 6112, Minneapolis, Minnesota 55424.
USA Electron Optics Div, 477 Riverside Ave, Medford, Mass 021.55.
Instron Corpn, 2500 Washington St, Dept L67, Canton, Mass 02021.
The National Engineering Laboratory, East Kilbride, Glasgow.
J. Goulder & Sons Ltd, Kirkheaton, Huddersfield.
David Brown Gear Industries Ltd, Park Gear Works, Huddersfield.
Carpenter Technology Corpn, Reading, Pa.
J. M. Leigh and M. G. Lay, Melbourne Research Laboratories, Melbourne, Victoria, Australia.

Honeywell Industrial Div, Honeywell Automation, Mail Station 448, Ft Washington, Pa 19034.
Carl Zeiss Inc, 444 Fifth Avenue, New York, NY 10018.
Radiography Markets Div, Eastman Kodak Co, Rochester, NY.
Magnaflux Ltd, 702 Tudor Estate, Abbey Rd, London NW 10.
Siemens A/G, Munich, W. Germany.
Avery-Devrison Ltd, Leeds.
W. & T Avery Ltd, Birmingham.
The Sheffield Testing Works Ltd, Sheffield.

Bibliography

'Analyzing Diesel Engine Failures', *Metal Progress*, April 1969, p 74

'Analyzing Fracture Characteristics by Electron Microscopy', *Metal Progress*, May 1969

'A New Theory on Stress-corrosion Cracking'. E. A. Gulbransen, *Materials in Design Engineering*, 47 no 1, p 150

A Photographic Study of the Origin and Development of Fatigue Fractures in Aircraft Structures. Min of Aviation, March 1961

Bibliography on the Manufacture, Properties and Testing of Steel Castings. Iron & Steel Inst, 1938

BSS 620, 'Testing Materials'

'Determining Residual Stresses Rapidly', *Metal Progress*, July 1969, p 88

'Determining the Toughness of Thin-walled Tubes'. V. J. Colangelo, *Metal Progress*, May 1969, p 117

Dictionary of Metallurgy, D. Birchon (1968)
Engineering Metals & Alloys, C. H. Samans (1949)
Engineering Metallurgy, R. A. Higgins (1968)
Engineering Science, J. C. Gwyther (1965)
Foundry Practice, W. H. Salmon & E. N. Simons (1966), pp 193–203
Information für die Tagespresse, no 8, 5.10.1966 no 9, 25.9.1969
Mechanics of Materials, F. Chorlton (1962)
Metallurgy, E. Gregory (1947)
Metallurgy of Welding, Brazing & Soldering, J. F. Lancaster (1969)
Metals Handbook, vol 1, 8th edn, American Society for Testing Materials
Procedure Handbook of Arc Welding Design & Practice, 6th edn, Lincoln Electric Co, 1940
Russian Metallurgy and Fuels, no 4, July–August 1960 (Eagle), p 82
'Short-term Tests Predict Long-term Creep-rupture data', *Metal Progress*, July 1969, p 86
Steel Castings Handbook, Steel Founders Society of America, 1950 edn
Steel Working Processes, Gregory & Simons (1964)
Strength of Materials, G. H. Ryder, 3rd edn (1967)
The Mechanical Testing of Metals & Alloys, P. F. Foster (1943)
Tooling, September 1970, p 26
Welding Methods and Technology, M. D. Jackson (Griffin, 1967)
Workshop Technology, Pt III, W. A. J. Chapman, 1966

Index

Acknowledgements, 196
AEG Test, 93
Automatic Non-destructive Tests, 191–2

Bars, Plates and Forms, 99–100
Beerbaum Test, 5
Bend Test, 13, 51–3
Bolts, 158–9
Brinell Test, 26–30, 189–90
British Standard 18; 1962, 22

Calibration, 14, 38–9, 93–7
Carbides, 173–4
Carbon Content, 23
Castings, 106–7
Closed Loop Electrohydraulic Tests, 183–4
Composite Metals, 174–5
Compression Tests, 13, 42–4, 78–81
Computerised Testing, 183
Core Loss Comparison, 132–4
Corrosion Tests, 135–8
Creep, 13, 55–8
Crow's 4-point Method, 23
Cupping, 91–2
Current Conduction, 134

Die Block, 22
Dilatometer, 69–77
Drop Weight Test, 88–9
Ductile Metals, 52–3
Ductility Tests, 91–3

Eberbach Test, 39
Elberack Test, 39
Electrical Strain Gauges, 180
Electrical Tests, 130
Electrodes, 159–60
Electron Diffraction, 117
Electron Fractography, 182
  Microscopy, 118–22, 187–8

Electronic Resistance Welding, 190
End Quench Test, 86
Engine Valves, 147
Explosion Bulge Test, 87

'Fastress' Analyser, 181
Fatigue, 14, 59–69, 189–90
Files, 166–91
Firth Hardometer, 36
Fluidity, 139–40
Fluorescent Tests, 127–30
Forged Parts, 105–6
Formability, 93
Fracture Toughness, 90–1
Future of Testing, 194–5

Gamma Radiography under Water, 151
Gears, 181–2

Hard and Brittle Metals, 53–5
Hardness, 14, 26–41, 152, 185–6
  Conversion Tables, 34–5
Herbert Pendulum Test, 40
High Temperature Test, 187
Hot Hardness Test, 33–6
Hydraulic Machine Testing, 183–4
Hydrogen Damage, 152–3

Impact, 47–51
In-process Control System, 188
Interference Microscopy, 192

Jovignot Test, 93

Kahn Navy Tear Test, 89–90
KWI Test, 93

Lead Corrosion Valve, 147–8
Large Scale Notch Toughness Test, 190–1

Long Term Creep Rupture Test, 181
Lorry Cab Tests, 186–7

Machinability, 140–1
Magnetic Analysis, 130–1
 Crack Detection, 125–7
Measuring Non-metallic Inclusions, 193–4
Metal Platings, 175–6
Modern Brinell Machine, 189–90
Modern Metallograph, 184
 Servo Potentiometer, 192
Molybdenum in White Iron, 151
Monotron Test, 39

Neutron Radiograph, 184
New Extensometer, 188
Nodular Iron, 157–8
Non-destructive Tests, 112–34, 191–2
Non-ferrous Metals, 100
Notched Bar Testpieces, 82–6
Notched Slow Bend Test, 89
NPL Test, 92

Penetrants for Magnetic Testing, 187
Permanent Magnets, 146
Portable Hardness Tester, 185–6
 Strain Indicator, 186
 Ultrasonic Testing Machine, 187
Proof Stress, 20

Radiography, 112–17, 151, 184, 193
Recently-developed Tests, 87–91
Reduction of Area, 22
Residual Stress, 142
Resistance to Oxidation, 141
Resonant Fatigue, 183
Rivets, 173

Rockwell Hardness, 36, 155
Roll Hardness, 152

Saw Tension, 171–3
Season Cracking, 143
Shear Testpieces, 81–2
Sheet Tests, 98–9
Shore Scleroscope, 40
Solder, 176–7
Spark Test, 163–6
Stainless Steel Quality, 153
Steel, Hydrogen Damage, 152–3
Stress Analysis, 177–9
Stress Corrosion, 143–4
Stresses in Aluminium, 188–9

Tensile Test, 14, 18–25
Testing, Equipment, 15, 19
 Miscellaneous Items, 110–11
 Object of, 14
Testpieces, 14–18, 75–86
 Preparation of, 14–18, 75–86
Tinplate, 177
Tool Steels, 161–3
Torsion, 44–7
Tubing, 100–3
Tukon, 39

Ultrasonic Testing, 122–5

Valve Elongation, 148
Valves, Engine, 147
Vickers Diamond Test, 30–3

Welds, 108–10
Wire, 103–5

Xeroradiography, 193